# Sungod the most probable theory of our time

# By Reg Griffin, was first published in 2003

## Book cover designed by Tracie Griffin
## UPDATED --- 2015
## Email: blueplunkett@gmail.com

<u>5-5-2000</u>

The "Year of Aquarius", the "Change In Direction", "The New Age of Learning".

Sorry Osama Bin Laden. What you did on September 11 was not for "The New Age". It was something from "The Old Age".

You will not be remembered as the man who began "The New Age". You will only go down in history as a mass murderer.

The "Old Prophecy" is nothing to do with human beings. It refers to the planets such as Earth and their lifecycles.

The whole world felt the Americans pain with 911, and yet another "prophecy" is misinterpreted.

---

## INTRODUCTION TO 20 YEAR PART TIME STUDY, 1980 -2,000

I need to make a statement at the beginning of this that I will repeat, later on. The statement will swing you into my line of thought, to enable you to follow, as you read further into this manuscript. The conclusion, will more than likely, be fact.
It is a thought followed through, and slowly gets more and more positive as more information seemed to just keep falling into place. I think it's a bit sad, as it took me more than 20 years to sort it out.

The statement is as follows:

"If you make a circle with your thumb and forefinger, where they meet we will call "summer".
Trace that circle with a finger from your other hand. We will call that circle "Earth's orbit".
As you continue to trace that circle with a finger, understand that each time you get to where your finger meets your thumb, is "1 year".
It is a very small circle, and if people lived on Earth as it traveled around the sun in such a small orbit, their year would pass very quickly.

They just might live for 967 years, as Methuselah did. We are not moving around the sun in an orbit of that size today.

-----------------------------

SUNGOD

EARTHS ICE AGE AND TROPICAL ORBIT

These are thoughts put down on paper to test the mind and the written word. Follow these thoughts to a conclusion, which that is more than likely to be fact.

"TIME"

An equation, that registers from the beginning of anything, to the end of everything.

THE MOST PROBABLE THEORY OF OUR TIME.
By Reg Griffin.

All myths around the world are saying the same thing although sometimes back to front. Myths and Legends, The Bible and all the ancient books, all balanced and fitted together tell this story. The late Malcolm McRae, a true friend of mine for more than 30 years, (who will be sadly missed), read my paper and gave me a book to read. Two days later I met up with him and said, "This book only proves me right". He replied, "I know".
And all the time I thought he was trying to prove me wrong. The book he gave me to read was" Fingerprints of the Gods" by Graeme Hancock. Most myths around the world are in this book and professors have their guess at what happens here on Earth and end up with more questions than answers.  I should have waited for this book as most things I have read in my 20-year study are in it.

--------------------------------

Logistic researcher, R J Griffin has revealed a surprising result from a 20-year study. For two years, part of his study has been sent around the world.
Australia. - South Africa. – Canada. - Sweden.

He has sent 1/3 part of his study to Universities, Museums, newspapers, radio and T.V. stations, NASSA and a lot of people who have just asked for a copy.
Read his papers, and decide for yourself, if he could be right.
His study on myths and the bible all tell us the same thing.
It took only interpretation to balance and place everything in line to see what it all means.
A mystery is only a mystery until all the facts are known, such as why and how the pyramids were built. The why and how list follows:

## THE GREAT WHY AND HOW LIST (Part 1)

1) For what purpose were the
   Egyptian pyramids really built?
2) Maori myths.
3) How did God give Joshua extra daylight?
4) What killed the dinosaurs?
5) What does the Earth's cycle do to the ozone layer?
6) What is "black rain"?
7) How is it that the position of the North and South Poles has changed?
8) Why are sea levels constantly changing?
9) Atlantis
10) Cataclysm

11) Lake Titicaca

12) Masada

13) The Egyptian seven years time, of plenty and seven years of famine.

14) The Aztecs

15) The Mayans and their myths

16) Here is why it really DID rain for 40 days and for 40nights

17) Why will fire rain from the sky?

18) Where might the planets in our Solar System originate, and how were they created?

19) What was it that totally dried up the Nile?

20) Why do all ancient books refer to Earth as "Earth 7"?

21) Why is today's time the safest period to be living in?

22) The world is definitely turning clockwise, isn't it?

23) Why could the South Island of NZ have been the North Island 10,000 years ago, (and vice versa)?

24) What halted the last ice age before it was able to get a real hold on Earth?

25) Why was it so critically imperative that certain ancient civilizations built their cities high in the mountains?

26) How is it that the sun has risen in the west and set in the east before today?

27) How is it that the natural life span of some people in history is reported to have been 900 years and more, whilst the natural lifespan of others at different times in the history of man, have trouble reaching 100 years.

28) How is it that time itself has changed? Why does it vary?

29) What is "faith"?

30) Floods

31) Ice on Mars

32) Small and major pole changes

33) UFO's

34) Lost Civilizations

35) Venus and Mars

36) Animals caught in ice at the poles

37) Equator changes

38) Pourangahua [MAORI CHIEF]

---

To follow, in a future publication, Reg's answers to:
THE GREAT WHY AND HOW LIST part 2

5

-Earths moving axis
-High-speed devices
-When things grow huge, (giants - and dinosaurs)
-The Big Shift
-Mini Ice Age
-The Sun in 2003
-Ozone and cloud
-Rotation Increase
-Sun and solar system
-King Hezekiah
-Planet strike
-The heavens fall from the sky
-The Heavens Were To-ing and Fro-ing
-Ancient observations and time
-Children of the future
-New Zealand South Island Rivers
-Freak waves
-The Plagues of Egypt
-GOD
-Displacement of Water
-How the Red Sea Parted For Moses
-Noah
-Daniel and the Lions
-One inch Scale
-The human Brain
-Theory on how the pyramids were built in Egyptian times.
-Earth's Ozone, ice-age, Floods, Earthquake, 7 years of Plenty, 7 years of Famine,       -
Earths rebalance, (pole change) and mass extinction, all in graph form.
-Moon and Earth at year 2000 and a surprising conclusion.
-Forecast of major problems of the future.

---

Part 1 Begins Here...

EARTH'S ICE AGE AND TROPICAL ORBIT 2000 YEAR CYCLE

AND THE PROBABLE EVENTS

[1,000 YEARS AWAY FROM THE SUN, THEN 1000 YEARS GOING
BACK TO THE SUN.]

Earth is very slowly being pulled into the sun, or to be correct, towards the sun. We will start from the last time of true Global Warming. (Roughly 800 TO 1000 YEARS AGO)

Earth will start global warming (meaning ice will melt back into the sea.)
Slightly closer to the sun and the world will turn tropical, just as it has before. Ice will keep melting into the sea and raising the world's water table, for a short time only.
[A note here- the water table is balanced by earths increased rotation spin, which forces the deep water to bulge and pull water away from the shore – all explained later on]  Some time later, plant life is at its most lush it has ever been for this cycle - a good time for dinosaurs.
As earth's orbit gets a bit closer it now gets too hot.
The sun evaporates water to cloud and is starting to evaporate water up higher and higher in the atmosphere which starts a major drought, as the clouds are unable to condense, (it's this huge volume of water hanging in the air which diffuses the sunlight, protecting the Earth from considerable damage).

The earth now starts to spin on its axis just that little bit faster and the orbit starts to pick up more speed, all due to the earth doing a smaller orbit and is able to complete the orbit in a shorter time .We are starting to get too close to the sun, and all of the planets in our Solar System are lining up, one behind the other.

------------------------------

I will pause here to explain another mystery before going any further. As the surface of the world is slowly losing all water, it is now unable to rain.
Let us concentrate on the people of the NILE.

The people of the Nile must have had plenty of water, a good crop growing and pasture for grain growing and farmland for animals - a self-supporting community.
Of course, they would have noticed the drought slowly taking hold - no rain or very little for some time and getting worse each day as the Nile river is much lower.

I think this would have happened slowly and worried people were forced to think that they needed to find water. Perhaps they already knew of the underground water table.

How did they get the water to the surface?

Possibly by noticing that evaporation takes place and forms clouds - as the clouds get heavier with water and more evaporation keeps adding to the clouds, they get too heavy with water and with no support, the water falls back to earth. A simple thing to notice.

Can they copy that simple, natural event? They set about doing just that.

Stone has very high water retention - they built a stone building with a wide base and the four sides tapered to the top in the centre. Sides and angles have to be correct for the building to work. They must have concentrated on the oasis, and then understood how it works.

If the water table is close to the surface, a small building would do. If the water table was deeper, a larger structure had to be built. Finally, the whole structure had to be covered with magnifying squares to intensify the heat. (People have been pulling off these glass type plates for their own use for a long time and a small earthquake shook off the unsupported plates that were left. I understand that there is only one pyramid left which has the full lot of plates remaining on one side only and this is now patrolled by guards at all times.)

Back to the structure and how it works. The intensified heat, at the correct angle, is forced underground. All four sides are pushing down heat to one point, by crossing over each other, underground.

The point of crossing is exactly the height of the building above. Four sides heating the ground underneath at the same point, = IS ON THE WATER TABLE. (Hold your hands in the shape of a pyramid in front of your face. Put your right hand slightly under the left. Then slide your right hand down your left to your wrist - take note where that movement stops. Now move your left hand down your right and you will see you have made a pyramid in reverse - an exact copy of the one above).

Was the plan for evaporation to take place and fill the pyramids with clouds that cannot escape (the building is locked - nothing can get out). The stone soaks up the evaporated water as if they were heavy clouds. Now too much water is in there and it simply falls out when the rocks cool and release the water = or did they study Masada, or more than likely, just simply copy how the Oasis is formed. More on the Pyramids later. Now - back to the earth's cycle.

The dangerous and nearly final part of the cycle starts to take place. To the extreme, water is being evaporated higher and higher into the atmosphere.

Earth's rotation is now getting too fast and it is interesting to note here that a Maori myth says just that. Have Maori been on this earth more than 2000 years? - That is how far back in time this happened, plus it has happened many times before. All kiwis have heard the myth that the sun was lassoed, and a famous Maori climbed up, struck at, and wounded the sun so that it now goes slowly across the sky. Maori then went on to survive the next terrible events - "obviously" ... Don't forget that as the rotation of the earth has been speeding up, so has that of all the other planets. Here, now, is the dangerous part of the cycle that I mentioned before. The orbits are more egg shaped when too close to the sun.

With the orbits getting smaller and smaller, the planets speed around the sun has been slowly increasing, until the point where the total mass of all nine planets speed and weight causes a slingshot effect. All planets around the sun move out with this slingshot effect.

The earth rolls and loses its axis in a slow deliberate wobble.
Another Maori myth slips in here - they tell us that north is south, and vice versa? Was earth traveling so fast at the tighter turn of the orbit that the world rolled over as it left that smaller orbit?

How much can myths tell us - is a myth only a myth because it is unbelievable?

The earth moves out from the sun, wobbling from side to side. From earth it would look as if the sun and heavens are zigzagging across the sky (another myth?).

What about all that water up in the atmosphere? Did it rain for forty days and forty nights? We are moving away from the sun, at great speed, so as the temperature is getting cooler, all of that water comes back to Earth.

What about this rolling motion? Did this action cause catastrophic movements of all earth's plates, resulting in violent earthquakes, volcanic eruptions, and the earth spitting lava? I hope our descendants are able to leave in space ships before all this happens.
Will we be in space at will in 1000 years? Were there many survivors last time as this part of Earth's cycle occurred? Obviously, some Maori survived, hence the myths. I have read many myths around the world which all support what I am saying about the Earth's cycle.

The volcanic dust has been forced up into the atmosphere, which the falling water brings back as black rain (another myth?). The sun is still pulling at the planets as they move out and away in wider and wider orbits. BUT the water is poisoned by volcanic dust.

Wherever the earths wobble is at that precise moment of swinging to that wider orbit, it is possible that the earth locks into a new axis. We are now in the fifth separate axis position in this cycle, today. So, again we see that this has happened before, so that's why I call it a cycle, - Earth's 2,000-year cycle.

Did somebody make it out there is space before this catastrophe happened?

Are they the people that keep coming back in space ships - that are seen all over the world?

I myself have seen such a ship of some sort. So did everybody in Helensville, (north of AUCKLAND, in NEW ZEALAND 45 years ago.) My mother called out at about 4.30am and woke us all. I thought cows had got into our garden, so I got up and went outside. No cows. It was cold and I was a bit annoyed when walking back inside. Then I noticed the whole family looking out mum's window. I too, joined in at peeking out and there in the sky and 200 yards from our house was a light as big as a house. (No sound.) For 3 seconds I watched as this light ever so slowly went down behind our cowshed, for about 5 seconds and then ever so slowly up to where I first saw it, and then it stopped. (I could not see any shape because the light was too bright). Then a knife-edge of light, full length across the night sky shot to the distance, and it was gone.

Everybody at school reported the same thing happening in their area.

Were they from earth at an earlier time? If they were, why have they preferred to avoid us?

Another Maori myth - Pourangahua was brought to New Zealand on a silver firebird. When the firebird, with this person sitting on its back took off, it burnt all surrounding scrub. He was away for so many days and when he came back he advised he knew of a place that was covered in bush and that they all should go there to live. [New Zealand] I am starting to believe that most myths have facts.

I was asked if I could reason why, when Joshua was in battle, he called God to lengthen daytime so that he could defeat his enemy. We must get two things together here to get a possible answer.

Why do the Maori suggest to us that we have got it wrong in saying "North and South have always been where it is today"? Having a hard look, at these two (myths) there is a possible answer that places both things to happen at the same time. The Earth would have to be spinning in opposite rotation. This brings in another myth from Egypt, ("Where the sun rises now, was at one time, the place where it "sets").

We come to a possible answer when the Earth is traveling too fast when approaching the tighter swing of the orbit. The sun holds Earth the strongest at the equator line. As the Earth is being pulled in one direction, and it's own speed, forcing it out in the opposite direction, the top side rolls down in the direction it is moving and rolls completely over as it leaves that small orbit.

Earth is now spinning in the opposite direction, daylight hours will be extended as the day is now starting again and the sun will be seen to go back across the sky from where it had come. This also makes the sun rise where it used to set - and the South Island (from now, till the next time it happens) will be the North Island, (deeper explanation further on).

FOOTNOTES:

1. As the earth's rotation picks up speed - the sun appears to travel faster across the sky.
2. When earth is leaving its small orbit, the earth is wobbling from side to side. This gives the zigzag effect of the sun across the sky. This also gives the appearance of the whole sky swinging out of order.
3. The further out earth moves, the colder it gets.
4. At one point, as earth moves away from the sun, animals would be caught at the new pole with a sudden temperature change.
5. As for the pyramids, the lower the water table, the larger the building has to be.
6. Stop drilling all through the pyramids - they are sterile buildings for a purpose. There is no gold or jewels in them!

---

I must go back to the UFO or UFO's that were seen in Helensville when I was a young teenager. I failed to describe something of real importance. We were looking out of my mothers open bedroom window when the light as big as a house came back up from behind our cowshed, (with no sound at all), it stopped for three seconds and then disappeared. It left a fluorescent white light which started where the UFO had been before it disappeared and stretched to the hills in the distance about eighty miles away. The light was only visible to us for a second, full length across the sky. If the light was visible over this great distance and turned on and off as if by light switch, then I assume it was traveling at least the speed of light.

I read about an American farmer who had a similar experience. The farmer saw a UFO land on his cattle farm and went to investigate. He found a dead cow.

He also describes the UFO and it's movements in the same way as my experience. Baffled by the death of the cow the farmer arranged an autopsy.

What they found was that vital organs such as the heart, liver and kidneys had been removed without incision wounds, and that internal wounds had been cauterized.

Somebody is well ahead of us in the field of medicine and surgery. Could someone have cauterized with a controlled laser and did they shrink the organs and remove them though the veins with blood?  I was talking to a lady about this and she said to me "That was not America, it was in Texas and I was 13 years old at that time, it was not only one cow, it was many cows. How is it that I was talking to a woman who corrected me? I was totally surprised.

Don't forget a famous AMERICAN quotation, "whatever the human mind is capable of thinking, we are capable of doing."

---

### Maori Myth

The sun and the fantail had a race across the sky and back. The sun cheated by only going part way across the sky, and then went back. The fantail, with his effort died on returning and was buried under a tree, which now flowers blood red.

That's how the Pohutukawa got its red flower, which is from the blood of the fantail.

The most interesting thing here is that the sun went part way across the sky and then back. One thousand years ago that did happen, when the earth rolled over and gave the appearance of the sun doing just that, and has done that many times before (a further explanation of this is given later on).

The fact that the Fantail is in this very important recorded myth proves that the Maori have been here in Aotearoa [New Zealand] for more than two thousand years as some have suggested.

---

How many times have I read this myth?

The sun and the Moon, FELL from the Sky.

It's a statement that won't go away. But it is true to the unknowing eye. Begin by yourself, facing north. The sun will rise from your right (East) and then the sun will set on your left, (West). Now we go back in time to when this "Myth" was first spoken. As I mentioned before, The Earth rolled over as it left the smaller orbit.

Now imagine, the sun rising in the west, then going back before getting very far across the sky at all.

Is this what they mean by the words: "The SUN fell from the Sky"?

The reverse of this on the other side of the globe (Earth) will have an extended night.

New Zealand was in a unique position when this rollover happened. The Maori people saw the sun go part way across the sky, and then go back the way it had come.
New Zealand just went through its winter on the top side of the globe, then rolled to the underside, (South Island now the North Island)

All the planets move out of orbit and too many in line on one side of the sun, allows them all to move farther out. I am sure a computer set-up could give a very good look at what happens. With a computer we can go forward and back in time, and we would be much more in the know.

New Zealand would circle round in this new orbit, to another winter, being now on the underside. But, as you have read earlier that's the least of everybody's problems. This has happened 170 times; to quote from books, (and it happens like clockwork). It takes 2000 years for each polarity change (Meaning earth rolls over). Quoting from writing in the pyramid (which says "they" have seen five suns die) = 10,000 years

$$2000 \times 170 = \text{Interesting}$$

---

## THE MOON FELL FROM THE SKY

This is also extremely interesting. If the moon were directly behind the Earth the moon would look like it stopped then went back, just like the sun. But if its position was anywhere else, it would look like it was falling in a direction it had never gone before, in an arc, across the sky. Then there is the problem of the earth rolling out of orbit, straight at the moon.

Could Earth hit the moon? Or does the moon back peddle fast enough to avoid the problem? Either way, I suspect there would be a violent reaction on Earth, lightening, sudden floods, terrible storms and high winds.

The plates would shift; the world would continue its 2000-year catastrophe all through the earth shifting its orbit.

The Most Dangerous Shift, The big shift of pole positions has to be when too many planets are in line on one side (causing a tug of war with the sun).

The earth will react with an over emphasized wobble, and be caught with a pole position, much more down the side of the earth, and the next (five or six)? Pole positions will be in close proximity as all planets have left their inline effect. We see again that a computer set-up would be a big help by telling us where all the planets are when the earth rolls, NEXT TIME.

-------------------------------------

### What killed the Dinosaurs?

The earth at full vegetation cycle was ideal for Dinosaurs. There will be plenty of food, and plenty of water, for all. Until, the earth was pulled too close to the sun. Water was evaporated off the surface of the earth and as it was too hot, the water could not condense and fall back. The meat eaters were the big winners here, as all they had to do was wait by the water holes. Rivers would dry up, and then water holes. All vegetation would be cut back and animals would die through lack of water. Meat eaters would take over the world. But the Dinosaurs are all on the way out. Food for meat eaters would start to get scarce, and being plenty of meat eaters, competition for food became a major effect, (the equator would be the first to run out of water).

Meat eaters being short of food would start attacking each other, and that is the beginning of the end. The only animals that would be left were small or too fast for Dinosaurs. Being smaller animals they would require less water to survive and would not require as much food. Birds and small animals would have access to water where Dinosaurs could not reach. I think this is when all Dinosaurs became extinct.

Then the world hits its 2000-year cycle of destruction and rebirth, (it begins again).
### Wide Orbit and small orbits

The reader does not see the reasons, and the follow-on, and constant questions force my hand. (Even though all questions are positive).

5/5/2000 is a very important date, as all planets have lined up with the wider swing of earths orbit around the sun. As the planet line-up has taken time (about 1000 years) earths rotation has been slowing down to just over 1,000 mph. Maximum slow down was this date, 5/5/2000, from now on with the planets breaking the in-line effect, earth and all our planets will start the very slow decrease in orbit and increase in rotation again, with the inline effect breaking up, the sun has regained control and will start pulling the planets back into smaller and smaller orbits.

I suspect the slow-down of earth's rotation has given us some readily identifiable problems, as well as unknown problems.

1. The ozone layer gets uneven around the earth, and so thin on one side that it breaks open, and a hole appears. This starts a hothouse effect, as unfiltered light and unwanted rays hit the earth, the rays that miss Earth bounce around the inside of the ozone layer, then eventually back to earth, causing unsettled weather and polar ice meltdown.
2. The deep water is able to flatten out and claim more of the coastal areas.
3. Earths orbit before the 5/5/2000 had been slowly increasing in diameter, taking longer to get from A to B, which forces us to add time to make up the difference and in the future, we will have to take off time, as the orbit is decreasing in diameter (referring to leap year).

-----------------------------------

## SMALL AND FASTER ORBIT

The sun, back in control after 5-5-2000, will keep pulling all planets closer and ever closer to the dangerous part of earth's cycle. Rotation increases and orbit decreases with obvious consequences. Deep water will bulge and pull all coastal water away from the coast. Only Islands in deep water, will slowly go under.

1. The whole earth will be hotter and more tropical. Earthquakes will be numerous as well as more rain. Further on in the future as all planets get closer to the sun, the in-line effect happens again, on earth's opposite side of orbit.
2. The last time in the spring – this time in the autumn. (Naturally this is from New Zealand's weather cycle, 5-5-2000 to 5-5-12000).
3. Being much closer to the sun, the rotation speed of the Earth increases because of the smaller orbit, and thus days and nights become only 3-4 hours long. It has also not rained for a very long time at the equator (plants have shut down completely) leaving a heavy cloud belt around earth. Earth cannot take its normal orbit at this speed and with

all other planets in-line, the combined speed and drag from the planets, forces earth to roll out of orbit.

4. Earth rocks from side to side as all planets leave their smaller orbits.
5. From earth (on the day side) it will seem that the sun is zigzagging across the sky.
6. On the dark side of earth the whole heavens would be rocking back and forth. All the plates on earth will move with this stress. Volcanoes would react in a violent manner, already in action because the Earth is too close to the sun...
7. The clouds would be able to condense and begin to rain heavily, bringing back the volcanic dust as black rain.
8. Every five or six axis change (we are in the fifth axis position now) there is a truly accurate line-up of planets, which makes earth react more violently. When the planets are not quite in line, there is a small axis change and how that happens is when the earth is rocking back and forth in its movement away from the sun, (earth is still being pulled at by the sun) it slows to a point where the sun stops the earth's rocking motion. Exactly then, the rocking axis was to one side, (=New Axis). A more exact planet line-up creates a much greater rocking motion of earth and the pole (or axis) is well down the side of earth when the sun regains control. Land mass which was nearer to the equator, will now be at the pole. You can imagine what will happen to living animals when that happens. The worst is still not over. We also have a new equator, which is in wrong shape. Earth is much wider around the equator line compared to around the earth pole to pole. The new equator has to expand to take its new shape and the old equator has to collapse (be sucked in as the new equator expands outwards by the rotation of earth - Atlantis comes to mind when you think about it).

There are a lot of places around earth that this violence can be seen. Along with Atlantis are Lake Titicaca and even Masada, show sudden lifts and collapse of the land mass. If we keep looking, we will keep finding places all round the earth where this has happened. Even landmass that has dropped has had landmass that was along side, fold and drop on top of itself. If that area was pushed well underground and if this folded area was well away from the Equator (which has now lost all vegetation) and was all green vegetation and heavy bush. Is this our future oil resource?

I know it is accepted that the world is now turning anti-clockwise. Standing at the "North Pole" the world is turning anti-clockwise. I am living in New Zealand, being under the equator line ("Down Under"). I prefer to face magnetic north with the sun rising from my right,- still facing north from under the equator the world is turning clockwise. (Take note, as, when the world turns over, I will be spinning in the opposite direction.)

When the Maori people saw the sun go part way across the sky, then go back to where it came from, they had just had their winter. What they saw was the world rolling over. To see it happen yourself, spin a ball (whatever direction you spin it, does not matter) view the spinning ball from underneath then look down on top. You yourself will see that from underneath, it is spinning in one direction and looking from above it is spinning in the opposite direction.

Now make a circle with your finger and thumb. Place a pen inside or point straight through the middle of that circle and turn it anti-clockwise, now keeping that movement turning, slowly tip that circle upside down - you will see it turning clockwise.

(Deeper Explanation)

1. The Egyptians tell us more than we care to accept such as a "seven year time of plenty" which I interpret as "tropical weather" - a time of rain and more rain. A time to grow plenty.
2. As the world is spinning faster, earthquakes around the whole globe have to be numerous. Liquid hot rock highly magnetic to the sun (until it is earthed out on the surface) and earth being much closer has this liquid rock surging world wide, and the smaller orbit is spinning earth faster and faster.
3. Too close to the sun the clouds cannot condense and "rain" as they should do. Again the Egyptians tell us something that we will not listen to ("seven years of famine" = no rain, no food = no river = every living thing will die without water. Masada is even today drawing water up to the pool at the top. All they had to do was to SEE how it worked then build a more efficient rock formation to draw water from the underground resource. It works like a hot wick burning and drawing its liquid. In this case hot rock draws water. Although all surface fresh water has gone, there is still plenty underground.

I will use this opportunity to explain something else to do with earths increase of speed in rotation. In earlier times those people in the know paid cash for gold at the equator, then had the cheek to sell the gold as close to the pole as possible.

I wonder if you have guessed what I'm about to tell you (OK - here we go). Gold at the equator is lighter (at the equator earth is traveling at a faster rotation).
At the pole, gold is heavier (somebody made a lot of money in earlier times until authorities woke up, or were made aware what was happening).

That knowledge tells me something about the pyramids and how they were built.

The world has picked up speed in rotation by roughly three times. Does that mean a nine-ton block of rock now is three times lighter, or should I say, it now is three tons? If they were, the amount of people (that we see painted on tomb walls), will be able to move them, they way they did. A man being twelve stone in today's times, would be four stone in weight at that time and there are people who lift three times more than I can.

4. The earth is rocking because in-line planets swinging around the sun are in fact pulling at earth, in a tug a war with the sun.
5. The Aztec Indians recorded the sun zigzagging across the sky = not so, as it was Earth rocking that gave that effect. 6. The Mayan Indians recorded that the "whole heavens were to-ing and fro-ing," which is the same thing as the sun zigzagging across the sky; only they are looking at the night sky.
6. Clouds being too close to the sun is unable to condense or form rain water. Only when Earth moves out and away from the sun are the clouds able to rain again as the Bible says "40 days and 40 nights it rained", (of course, this is in their reference to time, they had a 3-hour day and a 3-hour night).
7. There is a lot of races around the world reporting this deluge and the fact that the water was black. Very easy to explain when volcanoes are blowing black ash up into the heavy clouds that surround the world. I can keep adding myths to fit into this probable theory.

Maori Myth - The Moon God came down to Earth and took away a woman who held onto a palm tree with all her might. The palm tree and the woman (Rona) were taken back up into the sky, where she and the tree can still be seen, standing on the moon face, early morning through to sunrise. [This can only be seen on the New Zealand side of Earth as the moon does one full revolution per orbit.]

I find this interesting because, early into my writing about "Earth Ice Age and Tropical Orbit", I asked the question, "can Earth hit the moon when it rolls out of orbit?" The Earth must have come very close to the moon for the Maori people to record that fact in one of their myths.

Thoughts A

Taking a thought and following it can reveal surprises:

"Fire will Rain from the Sky"

This statement has been made in the Bible and my thoughts have gone out to find a possible answer. I have written with conviction, that the sun pulls all of our planets back in. When all planet orbits are getting closer, there is a slow in-line effect forming. The closer in they get, the more in-line they are. Too close and all planets are all roughly in-line and every so often they are perfectly in-line. All swinging around the sun together is resistance to the sun's control. The closer they get, the faster the planets circle, all pulling at one and the same spot.

Here is the situation that now is interesting. Planets swinging around the sun "as a unit," picking up more and more speed as the sun "pulls at them", sling shot themselves outwards away from the sun. Does this action pull part of the sun out, as all planets leave their smaller orbits? Flame and gas and liquid solids all follow the outward moving planets? IS THIS WHERE ALL OUR PLANETS COME FROM? - OUR OWN SUN?

Mercury and Venus will be hit first. Can they pick up all the debris before it gets to our own Earth? It might be possible. Taking this line of thought further - those two planets are the most recent and, if so, they will be more magnetic to any debris, which of course, means the less magnetic they get = the further out they are from the sun and the time scale keeps them all separate - which now creates another question; "did humans live on Mars before Earth was habitable?" -and, if so, "how long have humans been around?" We know that on the planet (Earth) we are still finding cities and human involvement well underground, which we call "lost civilizations". One of the latest being in Russia (Nexus magazine August - September 2001, which is quoted at being 5,000 to 7,000 years old) which I myself think is 1000 years ago. Mostly because 1000 years ago (and this being 2002) the world rolled over and created a new axis - this would account for major changes on the Earth, as far as water and global positioning of any well-established towns or cities. There is something else that I have just touched on - Earth's orbit from now, will get smaller and faster, four times faster.

Faster means just that. From summer to summer (one full year) will get shorter and shorter in time. A 50-year-old man in the future will be closing on 200 years old. The world will be full of people over 100 years old. I still think that future humans will be aware of what is going to happen to Earth and I am sure that we will simply get off, let the world go through its floods and earthquakes and when all done we will come back, straighten it all up and simply continue as before.

The worrying thing is, we must have information from past civilizations as the time scale is so great (2000 years) and being forced to start from scratch every time, we are forced to re-learn what was already learned in the past.

There must have been past civilizations that made it in space. The technology has to have progressed at one time in the past, as we ourselves have done, OR are we having outside help with our progress from other peoples who have left Earth at one time. If that is true, I am one who is not aware. We are told only what we are allowed to know. Possibly for our own good, I suspect.

The fact of knowing just could be too much for some races, which stick fanatically to firm belief (which stops any progress of knowledge in all directions) as we know from OUR past.

Back to the pyramids -there is a warning inside the pyramids that has been picked up by our scientists which ties in with what I have written and have attempted to explain with "Earth's Ice Age and Tropical Orbit". Let me tell you that I think that the Giza Pyramids are much, much older than our scientists believe. Twenty years ago, I picked up on the reason the dinosaurs died. It was not caused by a meteor and have been taking note ever since.

I suspected that the world left the sun and caused a major shutdown of plants and that was the reason they died out. In the year 2001 it all fell into place.

The reason for the plant shutdown was because we were too close to the sun and clouds could not do what they usually do, as explained earlier. No rain was the reason also that the Nile river dried up, as the world hit that part of its cycle. Our scientists have picked up that the three pyramids at Giza are set out in a particular star formation that will happen again in the future. They know the course of the stars so that course was set up on a computer and rolled forward. When that star position registered it was 16,000 years in the future.

It took 1000 years for planets to line up on the outer swing of planet orbits. Planet realignment, as they get too close to the sun, should take 1000 years again. I am sure that the smaller orbit was not taken into account = Time will be quicker in the future (not accounted for; 3 hour day = 3 hour night, three month year at the closest orbit to the sun)

Mayan calendar time for one solar year was 365.2420. Today's solar year= 365.2422. "WE HAVE SLOWED DOWN THAT MUCH SINCE MAYAN MADE THAT CALCULATION". There was no mistake on their part. As time is always changing.

Their system of calculation was extremely accurate, as agreed by all scientists. The fact that time is different, is the point here, as that calculation could have been anywhere in any of Earths outward moving cycles round the sun. Mayan being a very ancient society kept surviving each catastrophe, as did others.

We are now, of course, past the slow-down and now we are going back to the sun (as explained earlier). The Egyptians left us this message - "when the planets line up in the placement of the Giza Pyramids, standing inside the centre chamber of the larger pyramid and the planet Sirius can been seen through the vent on the right, the world will be in a GREAT catastrophe".

Scientists expect this catastrophe is going to happen, as such a catastrophe is seen to have happened in the past. Their question is "we know it happens but what causes it"? As this is still a mystery, I have put my thoughts to paper, which seem to be backed up by myths from around the world. Facts that are known to us all, [and our own Bible.]

-------------------------------------------------------------

## The First Time

## (Egyptian)

"Ra" the sun god looked down on earth and saw that things were not as they should be. He came down from the heavens and began to destroy his creation as well as to kill human beings. After so much destruction and killing, he tired and went back to the heavens dribbling ceaselessly. The effort was too much for "Ra", as he faded and died. When the new sun came to replace him, he was strong and bright and was given the name "shu". In time Shu did exactly the same thing that "Ra" did, and was replaced with another sun that was given the name "Geb".

There is a pattern here - or a cycle.

This is how I see that description of the first time.

1. The sun is coming down to earth - was the earth being pulled closer to the sun, which created a terrible tug of war - all in-line planets being pulled at, had picked up speed in orbit and just before slingshot speed, moved rapidly towards the sun.
2. Catastrophe on earth, destruction in all areas - finally the planets reached top speed (a reaction from the sun trying to pull them back in and the slingshot being the end result of "solid planet speed versus attraction to liquid solids (the sun).
3. "Ra", the sun going back to the heavens - happened when the planets left the sun and gave that appearance.
4. The dribbling from the sun as it went back to the heavens was caused by the planets leaving their small orbits and pulling parts of the sun behind them.
5. "Ra" did not get old then die, simply people could not see the sun as volcanic ash and heavy cloud blocked it out, and only when rain was formed and fell back to earth (as black rain) did the sky clear to reveal the sun again and the supposed NEW sun was given the name "Shu". This catastrophe happens on a cycle of 2000 years. A major catastrophe every time the planets are fully in-line. (Here is some guesswork on my part) We are told this next one is going to be the big one. By paperwork it probably is,

as the planet line-up on 5/5/2000, might be repeated in equal, when close to the sun. Nostradamus has predicted that the world ends in 3797. The Mayans predicted the "end of the world" in 2012, (which must only mean partial destruction of it), then again in 5125. The pole positions have changed four times in close proximity.

The first one, coming from somewhere else.(making the fifth position). It seems that Egypt had been on the pivot point of the changing equator each time, because not much change in world positioning seems to be recorded (immediately after the world rolled over). New Zealand was a pivot point the last time the world rolled over, as the Maoris recorded that the sun went part way across the sky, then went back the way it came. As the earth gets closer to the sun, Egypt will get tropical again, with all the extra rain.

Going back to the first time (Egyptian) it looks like "Geb" the sun god was the last of the sun gods as "Thoth" is quoted to be god in man. Thoth was the one quoted as mapping the earth and sky, and for all heavenly calculations. This is where I think the priests took over. Thoth being the first "God in man" must have been a method of getting control of the people, through the king himself. The priests needed a figurehead, and through the King God himself they applied all their knowledge, without Question.

Next God King was Osiris who introduced agriculture and a number of large-scale engineering, hydraulic works. (Maybe the pyramids) Is this still the all powerful priests still working without question? After all the God King has told Egyptians what to do, and their God King would take full glory of all new successful ventures, and maintain the God in man image. From Osiris to the next kings, god in human form ruled Egypt. Who was it that recognized that when water is heated, it climbs by simple expansion, and when we are a lot closer to the sun, hot rock would heat the underground water so that it climbs. Who was it that used hot rocks to hold water when expanded and then to release the water when the rocks cooled? The pyramids can't work as they should do now because we are "JUST NOT HOT ENOUGH". "We are too far away from the sun - that's why the mystery".

The Maori myths are more important to me than the others from around the world, so I feel that is what I should stick to, (being a New Zealander) as they are saying the same thing anyway.

## Maori Myth "1"

The sun and the fantail had a race across the sky and back. The sun cheated by going only part way across the sky, then went back. The fantail, with this effort, died on returning and was buried under a tree, which now flower's blood red. That's how the Pohutukawa got its red flower, which is from the blood of fantail.

The most interesting thing here is that the sun went part way across the sky and then back. One thousand years ago that did happen, when the earth rolled over it gave the sun the appearance of doing just that. (The long day and the long night)

The fact that the fantail is in this very important recorded fact proves that the Maori have been here in Aoteraroa, [New Zealand] for more than one thousand years.

## Maori Myth "2"

Pourangahua was brought to New Zealand on a silver firebird. When the firebird, with this person sitting on its back took off, it burnt all surrounding scrub. He was away for some time and when he came back he advised he knew of a place that was covered in bush and that they all should go there to live.

I am starting to believe that most myths are based on facts. Was this "silver fire bird" some sort of flying machine that they could only describe, as a silver firebird?

## Maori Myth "3"

There were complaints that the sun was going too fast across the sky so Maui and his brothers lassoed the sun, striking severe blows, wounding the sun to the extent it could not travel as fast across the sky, successfully lengthening the day. [A very interesting way to explain why the fast moving sun slowed down and lengthened the day]

## Maori Myth "4"

Rona was asked by her husband and sons to have a late meal ready for them in two days time, as they were going fishing. On that day she lay down for a quick nap. Suddenly she woke, and realized that the evening meal was not prepared, so she hurried down the bush path to the seashore with the moonlighting her way. As she was hurrying down the path the moon went behind a cloud and she tripped and fell heavily. She stood and shouted abuse at the moon. With that, the moon god came down from the heavens and grabbed hold of her. In her panic she grasped a palm tree; the moon god pulled her and the tree back to the heavens where she and the tree can still be seen on the moon face today (from early morning until sunrise).

This myth explains that when earth rolled out of its orbit it came close to colliding with the moon.

The Maori believe that the North Island is actually the South Island, but cannot successfully explain why. Their reasoning was that the South Pole is actually up and the North Pole is actually down.

This can be explained easily. "The world rolled over", makes the South Island the North Island, and visa versa. This event will occur again after 2000 years. [1000 years to go]

## Thoughts "1"

## EARTH "7"

Why do all ancient books refer to earth as earth "7"? Were people in the past advanced enough to know that Venus is the next planet to be in earth's orbit?

Starting count from Pluto, back towards the sun, earth is in seventh position (Venus would be number 8) A statement from the bible says, "When earth is finished, a new planet will come from the heavens". Is that planet Venus? Has man been on Mars? Did the human race come from Mars?

Scientists at the north and south poles found ice. Later the same scientists announced that the whole planet "Mars" is covered in ice [with rock and dust sitting on top and if that ice melted, every thing would sink, then the planet would be totally under water.] Mars is too far out from the sun to make water, so there is only one-way water could have been formed. Mars was at one time in earth's orbit! Earth is NOT the living planet; Earth is in the living ORBIT.

Mars in the future will come closer to earths orbit and a lot of ice WILL melt. Too far from the sun, earth goes through global warming which stops earth getting too cold. The ozone layer opens and lets in extra light, and that raises the temperature. Too close to the sun, water is evaporated into cloud which protects the earth from too much damage. When earth leaves the sun, clouds can condense and rain again.

## Thoughts "2"

We, who are living in this time, do not realize that this is the best and safest time to live on earth. There are fewer earthquakes, tidal waves, and in general less catastrophes, in our WHOLE lifetime. If earths orbit round the sun was down scaled to 8 inches round, being the distance we move out, from the sun and back, our whole life time in this scale would be 1/16th of an inch, or less. So from the time we are born to the time we die, we ourselves see very little change on earth. As far as earths cycle goes. 5/5/2000 being as far out as earth goes from the sun, Earth has now been put in this position (after planet in line) to return to the sun. In New Zealand, how it has rained and rained since that turn around.

Won't need that extra pipeline from the mighty Waikato to fill up our reservoirs, for Auckland community supplies. (When I was a child, it rained nearly every day in the winter. A farm I was on, just out of Dargaville, was really muddy in the winter). For those who don't know, just lately Auckland kept running out of water and bans were put on the use of hoses and we were given water saving ideas.

## Thoughts "3"

In my earlier writing I was trying to understand how anyone could build a pyramid without understanding of how it would lift water from the underground water table. It might have been their copy of what happens naturally to form rain. I think now, that the priests did a full-time study on Masada, which is a natural rock formation that lifts water. With their study complete, the priests know how to build a rock tower that would be more efficient than Masada. A wide base with four flat sides all leaning in to meet at the top, in the centre, and placed in the correct position would have two sides facing sunrise, four sides accepting the suns heat at noon, then two sides facing the sunset. This would work much better than a round tower. I think that the priests worked on pure logistics just as we do, and I can find nothing wrong with that. Companies, [now days,] have a section that deals in the very same thing. If problems are gathered into one section and full understanding of these problems are understood, logistics take over and the problems themselves give you the answer. A mystery is only a mystery until all the facts are known and understood. I think we can all relate to that. It starts when you realize that we are all working around a problem.

## Thoughts "4"

The Antarctic has been under ice for only 1000 years. When Earth rolled 1000 years ago the land mass that we know as Antarctica was somewhere else and much closer to the last equator. The roll over that Earth does every 2000 years is always in the same section of its orbit. The roll over has to be also at the same time of the year. If this theory is true it would be simple to find where the land mass Antarctica came from, and ended up at the South Pole. New Zealand's North Island would have been the South Island 1000 ago, if it was there at all. By the Maori myths, it was not, as Maui was supposed to have pulled up that land mass himself with his magic jawbone as a hook when he was fishing. Maybe he did hook something and as he was pulling in his line, the land mass was involved with an earthquake and was pushed up - Big Surprise.

## Thoughts "5"

The Mayan Indians must have witnessed what the Egyptians had seen. The Egyptians said the sun came down to Earth (his creation) to destroy and kill as many human beings as possible as things were not as they should be. To avoid that happening again Mayan gave the sun sacrifice in the thousands hoping to please the sun god enough, so that he didn't come down the Earth again. So many people died through just a total lack of understanding.

To gain knowledge has cost the human race dearly, it's about time it was understood what happens to our planet and all people stopped fighting, to pull knowledge and expertise into the field of space travel, which will have to be as soon as possible. Stepping forward with each generation would about do it in the time we have left ---- (major catastrophe in 1000 years). Do we subconsciously know that space travel will be our savior? Is that why there seems to be a rush to find out as much as possible? After all, this catastrophe has happened many times in the past.

## Thoughts "6"

The sun is the source of all positivity and all negativity to life, as we know it --once that was realized by people of the past, no wonder they looked at the sun and thought it was a god. The sun controls all vegetation, all the water and the very air that we breathe and also keeps the planet earth in the correct orbit, be it dangerous and wonderful at different times in its cycle around the sun. Revolution (rotation speed), being just over 1000 miles per hour and orbit speed 67,000 miles an hour and we don't feel a thing. As from the 5/5/2000 we will be increasing rotation and decreasing orbit, on our return journey towards the sun.

## Thoughts "7"

The sequence of this writing was as my thoughts came to me, and as everything was falling into place. I feel that I have to explain how I understood the Egyptian paper "The First Time". Initially, I did read the paper many times without fully understanding what it was trying to tell me. Then, whilst at the outset of writing my own paper, I still did not understand the Egyptian point of view 100%. Even when reading it yet again, my mind kept repeating the sentiment that it was all written in "Mumbo Jumbo". I was up to the sentence where "The sun god returned to the heavens dribbling ceaselessly", and I stopped reading right there, because something connected with the reading of that statement.

The sentence was saying something that I had written, but back to front. Standing on Earth and seeing the sun come towards them; it was easy to see their misunderstanding. They were not actually seeing what they thought they were seeing, i.e. it was not the sun coming down to the Earth. Rather, it was the earth moving toward the sun. What a thrill, as it suddenly made sense. This myth contains events of fact, just as the Maori myths do. And the sun returning to the heavens dribbling ceaselessly was the planets slingshot, away from the sun pulling out parts of the liquid sun behind them.

---------------------------------------------------------

Thoughts "8"

The Spinning Earth

I have touched on this matter earlier and there is a need to provide further explanation. From the date 5-5-2000, earth will be returning to the sun. The sun will be pulling all the planets closer in after that date, (5-5-2000). The planet inline starts to break up, which makes it easier for the sun to pull them back. This is because the combined focus of all the planets on a single spot in the sun is broken down as they move out of alignment. The sun controls Earths spin by the attraction of the lava, which is of course inside earth, creating its own magnetic field. As we get closer to the sun, we pick up rotational speed and the lava moves faster and faster. The build up of lava moves towards the sun and try's to stop at the closest point of the circle facing the sun, which creates a bigger build up behind the face, (or front) and as this build up is behind the front, the build up locks onto the Earth on the inside and then it pushes the front forward (or around) and forces a circular motion which spins the earth. The earth before 5-5-2000 was as far out from the sun as orbit goes, and as slow as it gets in orbit and spinning motion. This slow down of lava has allowed the outer circle of lava to cool. This results overall in less volcanic activity.

As Earth speeds up, the moving lava is able to reheat the outer cooled lava, through straight out friction, which will reactivate volcanoes all round earth. I am surprised to hear that scientists have already found that earth is picking up speed in the rotation (spin) of our planet. Scientists have noticed this in only three years.   [In three years, 2,000 to 2003, .026] Their explanation has been incorrect, as they say it is the expanding universe that is speeding earth up, (or it is only a guess). The speeding up of the earth in rotation will also speed up the jet stream of air around earth and that will spread the ozone gas more evenly which will then eventually close in the year 2,300.

The Earth, will also, in time, stop the wobble, as the speed up happens. Heavy weights (such as rock of which the pyramids were built) will be lighter and more manageable. Earlier in my writing I also mentioned that time will be shorter as our orbit gets smaller. One full year will get shorter and shorter (in actual time). This will result in people being one hundred, then two hundred (and so on) years old.

---------------------------

## Thoughts "9"

### Pyramids (Mica)

In the Tiahuanaco pyramid of the sun, underneath the floor, are two layers of sheet MICA. Mica has trace elements of different metals. This is very interesting, because Pyramids (as I have written earlier) were for lifting underground water by heating and forcing water to expand, and then climb upwards by simple pressure. Reading what metals are in mica forced me to ring the water board and ask what metals on this list are in our water today.
Some of these metals are in our water and some are not. They are interested in the metals that are not and were interested also in what I was studying. I explained about the future book and was asked to let them know when it was on the market.

---------------------------

## Thoughts "10"
### A Scenario

After the sun had pulled all planets back, and damage to earth by earthquake and floods had settled down to peace and quiet, I found myself a survivor. I first look for anybody else who might need help, or, is there no more survivors? I do find somebody and he is a friend of mine. In time we need food. I can see he has no gun or knife. We do see some animals who have survived. How do we kill the animals? A rock on the head? How do we preserve the skin for clothing? I do find a stick that I sharpen so as to use as a weapon. Do we sharpen a stone by chipping to make a cutting edge? I cannot make steel, so I can't make a knife or a rifle. I am sure that any bow I make will be useless. How do I make fire so we can cook the meat? In time, probably I would learn to perfect a friction-based system to start a fire. Everything we know has gone - no houses' T.V no clothing, no communications, no matches. All we see is tortured ground for as far as we can see in all directions. How do we survive? In time we find others. One of them is badly hurt. We are not doctors we cannot help. We need shelter we find a cave. One of us has to stay with the fire so it won't go out. We look like a bunch of "cave" men and woman. I'm sure you see the point. If we are unaware of Earths cycle and all this damage is done we will be forced to start again. If we are lucky, we will not have been left too close to the pole, with ice suddenly appearing it would make things even harder for any survivors.

Thoughts "11"

Mini Ice Age

In the 17th century, earth went through a mini ice age. That's what scientists called it. Earth was getting too far from the sun. Rotation was slowing down and the orbit was getting wider and wider. Earth moving a bit further out, stopped the mini ice age. The jet stream above earth slowed and was unable to spread the ozone gas evenly, resulting in ozone break-up which allowed extra light rays to get into the ozone circle, hit earth and pass earth, then bounce back to earth, [giving the effect of two suns,] which created the hothouse effect with all the extra light rays bouncing round inside the ozone circle. Planet line up with the sun on 5-5-2000 signaled the end of moving away and the return to the sun began.

I had wondered during research at what point did the ozone open. I had found no mention of this freezing before I wrote "Earths Ice Age and Tropical Orbit" as I was not looking for it. Then only a few days ago I saw a documentary on "Earths Mini Ice Age" in 17th century. Earlier, I had seen paintings of this ice age (in England) but did not tie this in as a problem that occurs as a part of the Earth cycle being described by this manuscript until now. We are going back to that orbit as we head back to the sun.

In the 23rd Century we can expect another freeze. Why I say 23rd century is because going away and slowing down, will take longer in time, as going away makes our orbit, wider. Coming back, earth is speeding up and our orbit is getting smaller and faster in time. We can expect the ozone to slowly close and to stay evenly spread. We then can expect the commonly frequent incidence of skin cancer that has been affecting people in recent times to fade.

The extra rays will once again be deflected away by the ozone layer. We can expect it to slowly get colder in the winter and hotter in the summer as our orbit gets smaller and faster, which will rip open the ozone hole on the left hand side as it is too thin. The ozone hole will end up as a slot reaching around the side of Earth. Summer for New Zealand will be extended and we will have a shorter winter. We will be still, too far from the sun and closer to the Mini Ice Age.

Knowledge gives us an amazing power. If we are expecting a mini ice age in the future, will we be able to control the effect on earth. Can we create a hothouse effect ourselves, all under control? We might put a disc in space to bounce back the suns rays, and direct it back to Earth during the daytime, so as not to effect the nighttime rest period that everything needs to survive.

----------------------------

Thoughts "12"

## MAYAN

Looking back, I seem to have skipped through explaining why the Mayan Indians sacrificed their own people in the name of their God (sun god). At the time of writing I was more concerned about the needless waste of life they suffered, through total misunderstanding - so now I hope to explain.

Everything was fine at the beginning - plenty of food, which they grew by using Pyramids to lift the underground water to the surface. Until, one day they all noticed the sun coming down closer to Earth. (Actually the Earth was closing in on the sun). At a certain position of the sun, the ground started to rumble, then RIP apart. The Earth erupted into waves and the roar of earthquake was deafening. The Mayans could not even hear themselves scream. The Earth was bucking and heaving, nobody could stand upright. The ground split then crashed back together, repeatedly. People were dying and the ground and crops were devastated.

The sun god had come to kill and destroy. Volcanoes blew themselves to new life and lava and smoke erupted into the sky. There was only one answer as far as Mayan people were concerned. The sun god had come for sacrifice. He was a blood thirsty God. A decision was made by the survivors to offer sacrifice to the sun so that the sun did not have to come to kill and destroy again. The slingshot that the planets did to escape the sun stopped the destruction. The Mayan continued to sacrifice to the sun as time went on, and the more important the issue, the more people were sacrificed. Did their society break up when the sun came down the kill and destroy again? Their sacrifice was already at MAXIMUM so they could do no more... [They are a very, very old race of people, along with the Egyptians and others around the world]

--------------------------

Thoughts "13"

Why did a lot of ancient civilizations build up high in the mountains?  Being ancient civilizations, they would have ancient writings from the past. The catastrophe that happens every 2,000 years would have been on record. The build up of this catastrophe would have been recognized, which means that they were expecting the earthquakes and floods - After the problems of the aftermath faded; there was no need to live high in the mountains anymore.

The community would have broken up afterward because they had free choice of anywhere to live, as everybody who had previously lived on lower ground would have perished. So the Mayan people were repopulated by the survivor's, along with those people in other ancient civilizations, such as the Egyptian's, Maoris, Aztec's, Inca's, and perhaps others who have perished since whom we do not know of in modern times.

---------------------------------------------

Thoughts "14"

## SURVIVORS CALLED GODS

Survivors of Earths catastrophe, (cycle) who have survived with technical people in different fields, first have to straighten up their own society and rebuild all the damage. In time they move out of their area in search of people who have lost all technicians and are surviving on a cannibal system - (that being the worst case which is reporting anyway). That would explain why the survivors thought that the technicians were Gods.

One of them had a weapon that when aimed at a person "would dismember you". They even have statues of that person standing upright with a weapon strapped to his side in a holster. He taught them how to grow food and taught them animal husbandry. So this man re-educated these people, then went on to find more survivors and to go on educating and civilizing.

This is reported all round the world by different peoples. What a kick start those men gave the survivors. No wonder they were given the name "Gods". Things were said and reported like; He came across the lake as if on a current but there was no current. Whatever he was on has to have been powered by something, doesn't it?

Quetzalcoatl was the name of one of these people who Re-civilized the rest of the world. Those men should be honored for the rest of time - FOREVER.

Going back to his weapon that he had - "when aimed at you, this weapon would dismember you". This does not sound like a six-shooter to me. A bullet will knock you down with a wound. Sounds more like some sort of ray gun. Dismember means to have pieces cut off you. There is also an ATOMIC wristwatch held by, whom, I don't know. Lots of things have been found, but we are not told - Probably for our own good I suppose (as I have written earlier).

I have waited for a long time, hoping to see that atomic watch on the market. They have only to copy it.

A rock had split open and it was found inside the rock, smashed. It went round the world a few times until somebody recognized what it was. "Where did you find that?" "It has not been invented yet." was his statement.

-------------------------------------------

Thoughts "15"

### STATEMENT FROM THE EGYPTIAN PAPERS

It seems silly for me, not to write down this statement, as it is so important to Earths Orbit.

"The sun is rising where it once set, FIVE times now."

Before slingshot speed, the earth's orbit around the sun is being pulled at by the outer planets weight and speed. When their speed reaches maximum, earth rolls completely OVER, then out of orbit, giving the appearance of the sun rising where it once set, because now earth is in a reverse spin. As I stated before, if you wish to see this happen, just spin a ball on a glass table top, look underneath the table at the spinning ball and you will see it spinning one way, then look down on top and you will see it is spinning the other way. For the Egyptians to record that fact, they have been involved for over 13,000 years. (As 2,000 years is the cycle). They also have to have been around longer than that to reach the educated science and to have seen and started recording fact. 10,000 years is where the changing pole is now known to be in close proximity = 4 times. That means that they have been recording these facts 2000 years before the major pole change. I hope you know how important this fact is.

The pole has had small changes FOUR TIMES, (the last four times that is). The fifth change being the first position had come from somewhere else.

That change was a major change and scientists know already that a full up society is under three kilometers of ICE, suddenly frozen, so it all should be still there. ANTARCTICA - and that came from much closer to the Equator, Egypt being the pivot point all the five times, allowed their area to escape damage.

-----------------------------------

## Thoughts "16"
## Akhenaton (Pharaoh)

In Akhenaton's young life, he had major problems. He had a birth deficiency that made him very unusual. His face was very long, as were his fingers and toes. He had very wide hips, an extended skull, and small breasts. These things had a very strong effect on the whole family. He was not included in affairs of state when the whole family was supposed to be there. Statues and paintings of him were not a revolutionary art style as many books suggest. (That is how he looked in real life). When sitting idle on the outside of his family, he must have seen the power that the priests had over his father. His father was supposed to be King, Pharaoh, God in man, yet he was told when, where, and how to do things by the priests.

Akhenaton must have vowed, when his turn came to be Pharaoh, that he would strip power from the priests. He did not want to be God in Man; he would go back to the one and only God - The Sun god. When he was made King, it must have been too hard to remove the priests as he walked out on his priests and built his own city in honor of the sun god.

Akhenaton did well in only one section of his responsibility as King. He gave everybody in his city a free and relaxed life, all worshipping the Sun god. But, he let Egypt's army run out of funds and outsiders quickly started taking over the outer limit, settlements of Egypt. Akhenaton and people who moved to his city all worshipping the sun god, did not sit well with the priests, and they could do nothing about it, they had no power over the king as he refused to be God in Man, Pharaoh. After 17 years Akhenaton and his wife died mysteriously, and the priests had his city pulled completely down, and then they tried to remove all trace of him.

A general, Horemheb was made Pharaoh next, and the priests moved to make him God in man, (to regain all their power,) and bring back their own Gods. Next to be Pharaoh was Tutankhamen at the age of about nine years old. (With the priests back in full control) Tutankhamen died in less than a decade and from what I read, died of head wounds. Deliberate or not, we will never know.

-------------------------------------

## Thoughts "17"
## Joshua

I must now explain in detail what I believed happened when Joshua prayed for extra hours to complete a battle.

1. All planets inline have picked up maximum speed round the sun. All pulling, at one small spot of the sun.
2. The sun being liquid cannot hold or pull the planets in any further, because of their speed.
3. The liquid sun is bulging outwards trying to hold the planets.
4. The planets have stretched the egg shaped orbit to an extreme, by their speed and weight.
5. All planets are pulling at this ever-increasing bulge in the sun.
6. The bulge suddenly bursts outwards.
7. This outward burst follows the outward moving planets.
8. This sudden release from the sun allows the planets to pull Earth out of orbit.
9. The two planets inside Earth's orbit also follow the outward rush.
10. Earth does not skid or slide out of orbit.
11. Earth rolls completely over when leaving its own orbit.
12. There are some parts of earth that escape the damage that other parts suffer.
13. This is where Joshua, and the reports around the world, said that the sun stopped. Joshua had to be standing on Earth in a specific spot, halfway between midday and evening when this rollover happened. (There is the polarity change that has happened 170 times in the past. All recorded in a Lava flow under the Atlantic Ocean.)
14. As the Earth rolls over it is still spinning. The further it rolls, the closer Joshua got to the side of the Earth. When the Earth has rolled over half way, Joshua was now on the side of the Earth and still seeing the sun. The Earth's spin is now totally vertical. If the Earth stayed in that position and kept the POLES facing AWAY FROM THE SUN, THE SUN WOULD APPEAR TO HAVE STOPPED FOR ALL TIME. As we know, it did not stop, so the Earth completed its roll over from half way, which then had Joshua on the other side of the equator, when the roll over is completed. The end result is extra daylight hours, as the Earth is now spinning in the opposite direction. [The day was ending. Now the day is BEGINING.]

-------------------------------------

## Thoughts "18"
### The Human Life span

Before the Genesis flood, in the past, people lived for 900 years. That means, Earth has completed orbit time at least 9 times faster. Orbit today is 67,000 MPH. A man one hundred years old today will be 900 years old in the very distant future. Summer-to-Summer one full year will be at least 9 times faster in time because of the smaller orbit.

No wonder time was different in the past, the question is, does the rotation of the Earth keep up with the Orbit time, the steady increase in the rotation will make the deep-water bulge, and generations of people will keep building cities that follow the receding shoreline.

There will be a lot of cities on high ground, under water today.

Earth's slower rotation is at 1,040 MPH and the deep water has flattened out, to claim land. As Earth picks up speed in rotation, there is a throw effect on heavy objects. The faster Earth spins, the lighter any object weighs. [Including, human beings, and solid rock.] Is this stage of the Earths cycle, the time when all things grow bigger and taller? Is this the time when men grow into giants? Just like the Dinosaurs of earlier times. This cycle happens every 2,000 years and this just might be "the length of a generation" that the bible mentions as the time that we have before meeting our God. Generation meaning, after the catastrophe the human race has to restart, then rebuild this NEW generation until the next catastrophe hits the last of this new generation and so on until we get it right. Last of the generation = 2,000 years after the flood.

----------------------------------

## Thoughts "19"
### Scale Down In Inches - Why Inches?

Earlier in writing I needed to scale down, the movement of earth's orbit, away and back to the sun into 32nds of an inch. 1/32 inches = 10 years. Therefore 1 inch = 30 years, 8 inches is 2560 years. This is a difference to the scientist's predictions that a major catastrophe will occur in 26,000 years from now. That calculation is worked out on today's orbit, which can't be right, as it is not allowing for the smaller orbit. Last catastrophe was 1000, which means the outer limit orbit was reached in 1000 years. I had to rework figures and came up with this scenario. The scale for time would be 1/32 = 100 years, x6 Inches is 1920 years and that is close to 2000 years as I can get using this scale to give us a picture.

1/32=10 years
1" =320 years
X6" =1920 years

We don't live long enough to see any change in orbit time. The last 14 years is the fastest movement towards the sun, so it's very hard to try and get exact.

------------------------------

Thoughts "20"

Speed and Time

Maori

"We had time only to rise in the morning, Cook breakfast, gather tools, and be at our fields when the sun would start setting." "Something had to be done." This could be the three-hour day of the myth, when Maui and his brothers attacked the sun so as to wound, and slow the suns pace across the sky.

Earth's Axis Rotation Speed

Year Axis Rotation Hours

| Year | Axis Rotation | Hours |
|------|---------------|-------|
| 2000 | 1040 mph | 24 |
| 4000 | 2080 mph | 18 |
| 8000 | 3120 mph | 12 |
| 12000 | 4160 mph | 06 |

(Here is the 3-hour day & 3 hour night)

--------------------------------------------

Thoughts "21"

Bible

Men lived for 900 years before the flood. Men live now for 70 to 80 years. This is what man has believed for a long time, and it has been unable to explain the apparently huge difference in life span between men of biblical times and men of the modern era. The confusion stems from the fact that until now we have been (quite wrongly) confusing our modern dimension of time with their old dimension of time. This is the wrong thing to do, because their day, if it were to be measured in our modern day hourly unit (for arguments sake), was much, much shorter.

Scientists are saying that the speed of the Earth, as it orbits the sun, is increasing. I am saying that the speed of the Earth's orbit around the sun is constant, i.e. it is always traveling at 67,000MPH. Scientists only think the speed of the Earth is increasing because they believe the length of one year (in time) is always the same. I am saying that as the Earth's orbit gets smaller, it gets closer to the sun; we complete our orbit in a shorter time. Speed is NOT the variable. The true variable is our perception of TIME.

Therefore: Constant factor = The Earth's speed, 67,000 MPH - Variable = an increasingly shorter year.

Human Years = Number of orbits in 1 year. (Measured in our time)
 80 Years       = Orbits the sun 1 time every year
480 Years       = Orbits the sun approximately 5 to 6 times every year
969 Years       = same speed, but Earth orbits the sun 10 to 12 times

-------------------------------------

Thoughts "22"

TIME IN MOTION

Earths wide, (big) and narrow, (small) orbit.

I will try another way of explaining Earth's change in time. We will use 2 cogs that mesh to help understand. One is a large cog that is driving a smaller cog and we will use the sun as the engine (or drive). The drive will be constant. If we start the drive the cogs will turn.

The large cog will turn the same speed as the smaller cog, as they are meshed, or have teeth with which to turn. The difference in the motion is that the large cog turns around one time and the smaller cog turns around 10 or more times. They are turning at the same speed. If they are not turning at the same speed, the teeth on the cogs would get out of time, then smash or strip. At the wider orbit the Earth circles the sun once. As the orbit gets smaller, there is less distance to travel so we get there earlier. Only time has changed. Hopefully you can now see why people of 1000 years ago lived for such a long time, as their year is too short in time. They have less distance to travel around the sun; however the smaller orbit does force the rotation speed to increase up to more than 4 times.

-------------------------------------------

Thoughts "23"

Water Down the Sinkhole and Weight

The closer any floating thing gets to a sinkhole, the faster it appears to spin around the hole. Exactly like the Earth in orbit around the sun, but there is no drain, there is just this pull from the sun and the increasing build up of speed is caused only by the smaller circle, and with the weight from the planets, the   liquid sun can not hold on, and the small area that the in-line planets are pulling at, bursts out, which releases the suns hold.

---------------------------------------------

Thoughts "24"

SEARCHING FOR DATES

2986 will be the beginning of the "seven years of plenty". (Tropical)

2993 Will be the beginning of "seven years of Famine". (No Rain).
At the year 3000 will be a repeat of the catastrophe that happened 1000 years ago,
(17th century) B.C.
Scientists have stated the time that they think this catastrophe will happen again, which is in 16,000 years.
5-5-2000 was the year that the planets lined up, and after that date the orbit will get smaller and smaller as earth goes back to the sun.

16,000 years is incorrect, as the smaller orbit was not taken into account. Their date was 16,000 because that equation was worked out in our own time. With the smaller and faster orbit taken into account, the year will be 12,000 and then 14,000 again it happens, on Earths 2,000 year cycle.

---------------------------------------------

Thoughts "25
SUN AND EARTH

The sun pulls at the lava inside the Earth. Daytime is always facing the sun, so the sun has always been pulling at Earth's lava. Earth is like a huge battery. North and South act like a positive and a negative. The lava churning around and the moon spinning around, all contribute to maintaining an electrical field. This field is emitted into space and it is that, which holds the moon in its orbit around the Earth. The moon's speed and the Earth's pull, keeps the moon in place. When rain clouds build up above Earth, the field is blocked and the field keeps building up in the clouds. When the cloud reaches a certain level of this electrical field, it earth's out, sending lightning down to Earth. If only we can harness that field and bleed the current so as to make use of it, before it discharges as lightning. There is another current emitting from the sun, all the way to Earth, (and all of the other planets too), which is pulling at the lava inside the Earth. This current is the most powerful of all until the planets pick up too much speed inline as a unit. Being in line and at speed, all planets are focused on the same spot in the sun, and the sun, (being liquid), cannot hold the planets' weight and speed. This pull from the sun must be similar to Earth's own gravity, and that is the energy we need for space travel. I am sure we will tap into this power source in the future. Earth has been in this outward moving orbit for 1000 years. Earth was once in the position that Mercury holds, then on to the position that Venus holds, and finally out to where we are today. (All planets originally come from the sun). This has all been recorded for us, but we won't SEE.

-------------------------------------------

Thoughts "26"

Earth - How a Planet is Formed, And, "THE BIG SHIFT!"

Our planets pick up too much speed around the sun, forcing the liquid sun to "bulge". When the speed of the planets reach MAXIMUM velocity, the bulging sun bursts outward, which releases the sun's hold on the planets. The separated bulge out of the sun follows the outward moving planets. The closest planet will be hit first, and if this planet is a small one, the direct hit will compress (this ends up as the core of the planet). The new planet will keep circling in orbit and keep being hit by the debris being released from the sun. Some of the debris will get past the new planet and hit planets, which are further out. Every time the planets come back to the sun and then leave it again, the closest planet picks up most of the debris from the sun. When this planet gets much bigger, the debris is able to hit OFF centre and that AREA is NOT so compressed. When it reaches a certain size and weight, all planets move out one position, (THE BIG SHIFT!) and another planet is started close to the sun. The weight of the planets has a lot to do with their ability to pull parts of the sun out as they leave.

-------------------------------------

Thoughts "27"
Earth's Distance from the Sun

To give me an idea of how close Earth gets to the sun, I rolled a 1 inch hose around a 6 foot diameter drum 5 times, each time on top of the hose from the previous revolution. Then I doubled that length, to get the ten times revolution. The scale is not right, but I wanted to know the length of hose required to do this. After laying the hose in a straight line it measured 60 feet. If the Earth was 1"in diameter, the Earth would have to close in 60 times its own width. At that point, the clouds cannot convert back into water, and the surface of the Earth would be losing even more water at a very fast rate through constant and continual evaporation, which is due to the enormous heat from the sun being in such close proximity to the Earth. We all know about this very severe drought because it is recorded in the landscape around us. The same is true of the catastrophic floods which occur when all planets leave the sun, caused by the cloud cooling off enough for the vapor to convert back to water and drop down to the Earth's surface again.

I know that mathematics can help us to work out these distances more accurately because the sun has a constant pull on the planets, and the changing orbit around the Earth adjusts "time", (see "Time in Motion", Thoughts 20).

---------------------------------------------

Thoughts "28"


DNA

If something happens here on earth, to affect the human race by threatening our survival (as far as environment and food goes), human DNA must restructure. Does the human mind control the DNA restructure? Scientists have seen it happen in our own lifetime. On an Island just off Japan, there are monkeys who were forced to swim for their food in winter. In our own lifetime we have seen the DNA change, as now; all those monkeys on that Island have their young born with web fingers. Come to think of it, of course that happens. It has happened to the fish, the birds, the lizards, and just about everything on earth has had a DNA restructure. That's how we got our DINOSAURS. It is called evolution; so don't expect races from the distant past to have the same D.N.A. we have now.


-----------------------------------

Thoughts "29"
The Bible

The True Religion When all the facts are known there will be no superstition, hocus pocus or mystery, because all questions can be answered with logistics and facts. We have a long way to go so let's get on with it. No religion is wrong. The apparent conflict between different religions is merely part of the process we must go through to discover the truth. Let no man die because of his own faith. We are almost there, where we are at now is surely an improvement on where we've come from. We must keep going. Ancient beliefs still rule religion.

God does not come down from the heavens to kill and destroy. In the "past", God came down to kill and destroy. Please get it right! Their God was the Sun god, not the God that we refer to. Our God IS NOT a jealous God, he is beyond such things. We are here on Earth, able to exercise our own "Free Will", with NO INTERFERENCE from God. We are here to learn how to distinguish between good and bad for ourselves, so that we will be able to make decisions of benefit to mankind. The alternative is to make decisions that may seem of short-term benefit to us now as groups or individuals, but which actually have negative long- term consequences for mankind. This is the double-edged sword of FREE WILL.

Too many evil things are happening here on Earth right now, and in the past to prove that God will not interfere. It is no worse now than it has been in the past. Terrible things have happened on this Earth by the hand of man. Things too terrible for me to write down as emotions and shame make it nearly impossible to do so, because pouring them out without thought is possible. But to write them down forces one to dwell.

I, for one, believe in God. My faith is always with me. FAITH can move mountains. FAITH and PRAYER focus's the mind in order that you can help yourself to reach whichever must be achieved. That is where the win is. God IS helping you through your own FAITH!

Get rid of the old belief. God does not kill and destroy. He is beyond jealousy, and any other human frailty you wish to name or describe. He is a loving God. That is why we are still here.

If we cannot get it right before "two thousand years", meaning, "all for one and one for all", we are all surely going to die or nearly all be killed in the next great catastrophe. Only by working together, all of us from all over the world, towards a common cause such as the continuation of the human race, can any of us survive.

Consider the spoken word of God when he said; "Go forth and multiply" He did not mean, "Have lots of babies". He was commanding you to multiply yourselves. Your children are what you make them. Some of you are successful with them, some are not. If some are unluckily from "the bad seed", they will always be a problem, to whoever the parent is. But they are still what you made them. Let's take another step. "Multiply yourselves". All men then, are "Adam".

All women are "Eve". Now we can debate and learn. The more people involved, the more debate, and hopefully, the more we can all learn.

If we don't learn, then we are in trouble, as "the last of the generation" before the great catastrophe will nearly all die. The survivors will have to start the "next generation".
If the last of this generation can get themselves sorted and make it into space before the next great catastrophe, and fully understand DNA to the point where we cannot die, then have we arrived at the place we call "heaven"?

Is that when we can introduce ourselves to our "God"?

We would leave behind a planet that is too close to the sun (Earth's cycle). Reports from back at that time mention "the very Earth itself was on fire". That sounds like man's version of "hell", reaching out to warn us.

Ancient civilizations thought that the sun went across the sky, down into a hole and traveled through a tunnel under-ground, back to the opposite side, then back up through a hole and across the sky again. The sun going across the sky by day, then, underground by night. People, who know the bible inside and out, have been asked this question repeatedly. "How did people before the flood, live for nine hundred years?"

Please don't answer that question by saying that "Time was different then, the sun was going too fast across the sky." There is only one way that the sun can go too fast across the sky "THE WORLD PICKED UP SPEED IN REVOLUTION AND ORBIT" Do not use Ancient beliefs to try and explain that. GET IT RIGHT. As the world and all the planets leave the sun, our orbit gets wider and wider, making our year longer and longer. People, who lived on Earth, when on a much smaller orbit around the sun, have a very quick year and appear to live longer. When all the facts are known, there is no mystery.

---------------------------------------

Thoughts "30"
Them and us. Why are we here?

Who are they? They are bigger, stronger and faster. They are soldiers, programmed to protect us. Programmed to seek and kill to protect us. I find that information staggering. I am talking about reproduction. [The male sperm] I watched a documentary that used a camera to actually see what happens to the male sperm when first entering a female. The very first of the sperm are soldiers. They spread out looking for anything that should not be there. When something is found, they all attack as a unit, leaving the others to go where they should be going, and taking no notice of the battle as they rush past. This was shown to the viewer as a known fact. As something that is accepted by all those in the know. Did those in the know take that information a step further as I did, as the, "what if" took over? What if one of the soldiers got caught up in the rush of the others, trying to get where they are all going? Fight as he might, he is alone. The mass carries him away, and does the need to survive then take over? He is stronger, bigger and faster, he can easily outpace the mass. What is the end result? Is he still programmed to seek and kill? Is that why some humans kill without any remorse or guilt? If this is possible, it must have been designed that way.

A killer is placed amongst us on purpose. There is only one reason that makes sense, and that is, to make sure we learn right from wrong, good from bad. Is that why we are here? When asked "Why did you kill all those people and bury them under your house? His answer was "I don't know".

Was he a solder sperm at the very beginning? Obeying his programming? Does a soldier sperm, which works out what he is, become a mercenary and actually gets paid to kill? Is he too, obeying his early programming, and in a way still thinks he is protecting someone? All religions have their own faith.

All religions teach right from wrong. Maybe we are not here to find what religion is correct. Maybe we are here to learn right from wrong which makes all religions correct, in their own faith, which brings to mind "Faith can move mountains".

With some afterthought, I would like to take this subject a bit further.
As time goes on, the human male produces less and less solider sperm, to a point that there are only one, or two soldiers, then none at all. "The meek shall inherit the earth".

That statement will then be a matter of cause. As the human mind develops and concentrates on getting things done without violence, the soldier sperm is less called for.

What I am saying is, the human mind, at this stage will still produce soldier sperm and if this thought runs close to being fact, it will be picked up on individual D.N.A. at birth or maybe before birth.

This is only a "What If" scenario, a thought followed through, it seems to explain a lot but of cause, no medical evidence (yet).

---------------------------------------

## Thoughts "31"

### EXPERIENCE

A Self taught man is a thinker. Experts forget to think. Experts are taught a method so don't expect an expert to solve a problem outside the method he has been taught. Believe in what you have to believe, but don't close your mind. If a person believes he knows everything, he closes his mind, which stops him learning anymore (very unfortunate).

---------------------------------------

## Thoughts "32"

### Memories

What happens to Memory's, when one passes away?
Memory's that are such a delight, it brings tears to one's eyes.
Memories that fill ones heart with pride
Make one burst out in laughter
Make one so sad
Make one feel shame
Make one full of admiration
Make one angry
Make one feel regret
Make one feel a terrible loss
Make one feel an all-consuming LOVE
One should avoid the feeling of hate. Hate should be a regret that might be rectified, and if it can't be rectified, that is a loss.

44

Memories are emotions.

Is one's emotion's, our soul? And is it that, which survives the death of, our body? Does soul and spirit mean the same thing? Or, does the soul live within the spirit?

Still searching...


REG GRIFFIN


BY LOGISTIC RESEARCHER
R.  J.  GRIFFIN
[[YEAR 2.000]]

FOLLOW THESE THOUGHTS ON PAGES, TO A CONCLUSION
THAT IS MORE THAN LIKELY, TO BE FACT.


------------------------------------------------------------------


EARTHS OZONE, ICE AGE, AND   TROPICAL ORBIT


PART TWO

Earths moving axis and high-speed devices

Planet alignment with the sun happens every 2000 years, once when close to the sun, then once again when at the extreme orbit away from the sun. EARTH has had 5 minor pole changers all in close proximity.  How many more pole changes before a major change? It is crust displacement that forces a pole change .The last MAJOR Pole change was 1000 thousand years ago.  Can we use a computer to go back 1000 years in time to look at the dry land on the surface of earth, then roll it forward and watch the land movement around the entire globe, move to another position and force the spin of earth to get out of balance?

I think THAT, is the reason for the small pole changes. Does the sun speed up earth simply to refocus the axis and stop the wobble that earth has? It seems that way to me. At this time, earth is out of balance, as WE are wobbling in rotation, and of cause, earth just went through its slowest rotation speed on the date 5-5-2000.  Because of earths round shape, we should not be wobbling, being extended round the equator should let us spin quite freely at slow speed, unless landmass moves and forces otherwise. The year 2-300 will be dangerous for human population. We will be back at the same position as we were in the year 1700. [From 1700, earth has moved out 300 years to its outer limit in orbit (5/5/2000)]. From now onwards earth will be going back to the sun, picking up speed, because of the smaller orbit. Mini ice age here we come, we are going back to the same ICE -AGE as in the 1700's.

=============================

Does the earth let us know in advance, the amount of change the pole position will have? I still am sure that in 1000 years, the best place to be is in space. Let the damage (earthquakes and floods) occur, then come back and straighten it all up again. But in saying that, how many of the generations before us made it in space? (One generation being only 2000 years) Maybe they have found a better place to live, and have not bothered to come back. We know that someone keeps coming back for a look see, as they were seen all over the world, but not lately. I wonder why? There just might be a very good reason. Something ELSE is having a look see…

Someone was filming an activity [[men were in South America jumping down a very deep hole with a parachute, then coming back up in a makeshift lift and jumping again]] when something flashed past the camera and that something, was thought to be a bird at high speed .A little time past, when he decided to slow the film, to try and pick up clearly. What type of bird? To his surprise, it was something that has never been seen before [or so he thought]. It was a white rod about two and a half feet long with flexible sheet sides attached, that were full length of the rod either and were moving like a wave at high speed. The rod itself was the size of a small piece of broom handle and the flexible wave on both sides of this rod was about four inches wide.

He had no idea what they were, so he decided to put his film on internet with his address and such, asking if anybody knows what these rods are .He was flooded with film from a lot of people, all asking the same question. To date we still don't know what they are. I have heard that they might be insects, but I don't think so. If they are, we better get hold of one and copy their method of high speed travel, as they are so fast we cannot see them with the naked eye. I think, they are a device that travels at high speed, so they can be undetected as they film or listen to whatever. They are even out in space, as a film shows them whizzing about up there. If they are devices, the technology is far ahead of ours and one can only wonder, as we have not even seen one up close, dead, alive or damaged.

-------------------------------------------------

## WHEN THINGS GROW HUGE

Let's look at the dinosaurs. Huge and heavy? Huge yes, but not heavy as we know heavy. Living in our time, their bulk would make moving around, a major problem. The further out earth moves from the sun, the heavier any object becomes. The earth slows in its rotation, which increases gravity. Year 5/5/2000 is as heavy as any object becomes, but as earth closes back to the sun, things change, time and weight will not stay the same. As earth picks up rotation speed and the earth is completing its orbit in a smaller circle, TIME gets QUICKER and WEIGHT gets LIGHTER. The most vicious animal on earth to reach a giant size has to be a rat. When I was younger I read in the paper that students found on an island, skeletons of rats, THREE FEET HIGH. There is only one way we could defeat such animals and that would be a united band of men working together with an annihilation plan of attack. Obviously, one on one would be fairly impossible. The time that those rats were so huge, man was what we call giants as well. As we all know about the giant called Goliath. He was supposed to have been one of the last of his kind, and a city of giants, [1000 years ago] were all killed in the catastrophe of earthquake and rock and fire falling from the sky. BY THE WAY, our Christian god did not kill these people and totally wipe out their civilization. The so-called sun god was part blame. Earth was too close to the SUN, ALL PLANETS WENT THROUGH THE SLINGSHOT EFFECT AND PULLED OUT LIQUID ROCK FROM THE SUN, and HITTING ALL OUR PLANETS ON THE WAY OUT. As explained, in part one.

--------------------------------------------------

## MINI ICE AGE

## THE BIG SHIFT

In the year 17.00, the earth went through an ice age and just as quickly had a complete reversal in the weather and thawed out. The earth just spent 1000 years moving out away from the sun - [[1000 years away from the sun and 1000 years back to the sun, which equals 2000 years and is the GENERATION THAT THE BIBLE HAS EVERYBODY WONDERING WHAT THAT ACTUALY MEANS [[2000 years equals ONE generation]]. Lets go back in reverse 1000 years ago to the beginning of this generation [back to the sun] to the last catastrophe. Follow back out, earths last movement back to the sun [still in reverse] back out 1700 years, [in total back in reverse 1700 years] earth had an ice age there. NOW come back in to 1000 years ago, there was an ice age there as well. [[More on that ice age is explained further on]] Keep coming back to the sun, to the catastrophe of earthquake and floods, [[to the beginning of this generation]] and come out 700, which will be in OUR time, the 17th century. The earth was too far out from the sun and froze for a short while, the earth had slowed in rotation speed, [[Being too far from the sun]] which now, at that slower speed, is unable to spread the OZONE properly and most of the ozone went to one side of the earth and on the opposite side got so thin that a hole ripped open. Extra light burst in and melted the ice back to the pole. This has all happened before, and will happen again. It looks like the planets pick up debris from the sun, when moving away, and more weight is added, which carries the planets further out, away from the sun. It looks like a slow process and not the sudden BIG SHIFT that I thought. More weight, faster speed out from the sun, all the planets move.

WARNINGS from the past, might mean the BIG SHIFT, is either the big shift of the pole positions, or ALL planets move out one and that IS a big shift. ----- GOING BACK to the Ice-ages, puts earths cycle on the way back to sun, so that means we [[from the year 2.000]] are doing the same thing NOW -----MINI ICE-AGE HERE WE COME in approximately the year 2.300, then again when the orbit is much smaller, ----- in THIS generations time.]

-------------------------------------------------------

OZONE AND SUN   2.003

It is not the going back to the sun, where earth's demise is. When earth goes in, we slingshot out. It is the going out that is the big worry. When earth goes out too far it will leave the orbit, that life as we know it, cannot survive.

Earth in orbit closest to the sun, when the volcanic, drought then flood problems, are all over, Earth just picks itself up and regenerates as it has done many times in the past. Human survivors restart the new generation…

Search as I may, I cannot find any reference to the hothouse effect that we are having now, at any time in the past. It must have happened many times in the past because THE ICE AGE BACK IN THE 1700s, 300 years ago, was stopped by the ozone opening, and then for 300 years the earth moved out to the extreme orbit, and now we have to go back 300 years. That means 600 YEARS and the ozone has been open all that time, until earth picks up enough speed to close the ozone hole in the year 2.300.

Scientists know what happened the last time earth went back to the sun. So our weather pattern will be predictable [they know of the floods and earthquakes, don't they?] I still wonder, how many of the last generations have got themselves into space before the catastrophe hit earth. There is, records in writing, stating ----- THERE WAS ROWS AND ROWS OF CHARIOTS WHICH TAKE THE GODS TO THE HEAVENS. That could only mean, someone is getting into something, and flying into space and these people would have to have been much further advanced, because the people watching call these flying objects

CHARIOTS --. WHO WERE THE PEOPLE FLYING AROUND IN THESE SO CALLED CHARIOTS?


## SUN

During the Falkland war a phenomenon happened, and it was obvious that it was unknown at the time. England had built her warships in aluminum. Nobody knew that if weapons of war were fired at the ships, a certain temperature would be reached and actually start the aluminum burning. The heat that builds is enough to heat the surrounding area, hot enough to keep it burning. It seems to me that our own sun is doing the same on a much bigger scale. Was our solar system formed when two solar systems collided? Were they two very old solar systems with planets PACKED around their own sun? Did all the planets unbalance their whole system and cause two systems to collide. Is that why there is a hole in our universe? Was nearly the whole of two solar systems crunched to a big mass? And leave the hole that we see today, as there are no planets there now. (That was the closest solar system to ours.) Did they reach critical temperature, and start burning? Earth, at this time is traveling at 67,000 M.P.H. in orbit.

You can easily imagine a very old solar system tightly packed with planets, and all traveling at that speed, there would be a mighty big extended BANG --. Then, there is what is left of the two suns to collide as well, all into the same mass. It's like a rebirth of a new solar system. TWO INDIVIDUALS JOIN TO MAKE A NEW ONE --- [[I have heard of that somewhere before]] STARTING WITH THIS NEW SUN, WHICH IS, OUR SUN, and is still burning with a very, very, long, LONG, WAY TO GO. This solar system has to have all the planets pulling our sun to bits, just like all other solar systems until the sun [[at this stage would be very small]] is surrounded with planets full circle, many times around, then the sun loses control, and all in one mass are attracted magnetically to the closest solar system and collide into one complete mass and the friction involved, all burst into flames. = NEW SUN.

-----------------------------------------------------

## OZONE AND CLOUD

When earth is in orbit, as close as we get to the sun, what saves us from the expected heat is WATER. The sun has evaporated so much water into the air, and being so close to the sun, it is too hot and unable to rain. All the water is in cloud form and small droplets of water hold themselves around small dust particles. These small drops of water in the millions are defusing the sunlight by splitting and spreading any light that gets through. HERE is where both OZONE AND WATER together act as an efficient sunlight barrier.

49

There are reports from way back then---In some places around earth it is very HOT ---The equator is closest to the sun. --- STATEMENTS LIKE ----It is so hot that trees are bursting into flames. It is so hot that the very ground itself is burning ------- WITH ALL THAT HEAT, PEOPLE STILL SURVIVED. --- ALL THAT TIME IT HAS NOT RAINED. [7 years]. After the heat and earthquakes, all the planets in line pull away, (with all that extra speed in the small orbit.) letting the heavy cloud surrounding the earth, cool down enough to convert back to water --- FORTY DAYS AND NIGHTS IT RAINED. They only had a three hour day and night there, so that statement was from their time, NOT OURS. The orbit was too small, and time was too fast.

## ADDED NOTE [2006]

Word has trickled down from those who have data on the universe. They also have data on what speed that the universe is moving out and away from us. (5/5/2000) Earth has reversed its outward moving orbit. We are going back to our own sun. Anything in the universe that is moving away from us would have a recorded speed. Earth is now in reverse movement, picking up speed, as the orbit is getting smaller, we take less time to complete the smaller orbit. Only our solar system is picking up speed, the universe is not. The universe will appear to be moving away, gaining speed, but it is earths reverse movement that gives that appearance.

---

## MOON AND EARTH

The moons speed in orbit is faster than earth's rotation speed. Moon orbit speed is 2.300 mph. Earth's rotation speed is 1.040 mph. The moon is trying to leave earth with its speed. The earth is pulling or holding the moon from moving out any further by positive and negative force. With this speed and hold, the moon stays in place with a balance between the two.

---

## ROTATION INCREASE

Scientists have given out information on the earth's increase in rotation speed. In three years the earth's rotation has increased by point 26%. (From 2.000 to 2.003). As you know by now, that I have been expecting this increase in rotation. Rotation is a day-by-day event so it was picked up early. Orbit is a year-by-year event, and that is still to be picked up by our scientists. You also know that I am expecting the year to be faster as well, because the orbit circle is getting smaller

## KING HEZEKIAH

The king was very ill and had visitors, one being a prophet, told him that he was about to die. The king got upset and started to sob, hearing this the prophet turned back to him and told him, "God has just spoken to me and told me that he has heard you weeping and to tell you that if you would become a virtuous man, he would grant you another fourteen years."

The king said straight back. "If you are speaking to God, ask him to give me a sign". Whether it was the kings own shadow that moved from in front of him, to behind him, or the shadow of a sundial that changed position it was still the same thing that happened to JOSHUA [see JOSHUA, part one. All shadows move to the opposite side. The long day and the night --The sun stopped] every 2.000 years the world rolls over and gives the appearance of the sun stopping, then moving back to the opposite side, Now the sun rises where it used to set. "The long day and the long night."

-------------------------------------------------

## PLANET STRIKE

The so-called meteors strike on all our planets happens to be roughly in the same area of each planet. I think, that when all planets are close in on the sun, it will show that all the damage will be facing the sun. Planets will face the sun one after the other as debris that the planets pulled out of the sun, hurls itself outwards at high speed. As far as the Mayan prediction that the world will end in 2012, there is a possible explanation as to how and why that prediction was forecast. -- Way back in the past, a certain asteroid whizzed through our solar system. -- A very long time later, that same asteroid whizzed through again. How many times did that asteroid pass through before that prediction was made? Every time it went through, was it a bit closer to earth? Don't forget that the Mayan was deeply involved with the heavens and they were a very ancient people who kept on surviving the end generation catastrophe, with records of that event. That asteroid is coming now [[scientists are watching it]] and will be here in 2012, BUT, EARTH WILL BE IN A SMALLER ORBIT IN THE YEAR 2012. --- Are we moving away from this asteroid or are we moving closer to where this possible collision is predicted to happen---. THAT IS THE ONLY DANGER TO THREATEN EARTH IN THE YEAR 2012 THAT I COULD FIND, AND I HAD TO WAIT FOR THAT INFORMATION TO COME FROM OUR SCIENTISTS.

-------------------------------------------------

## THE HEAVENS FELL FROM THE SKY

The heavens fell from the sky, is a statement from a lot of cultures around the world. [I wonder if you have already figured this one out, by reading earlier] If a person, or a group of people were standing on earth in early evening, full night or early morning and happened to be looking up at the heavens when the earth rolled over, [THAT WOULD BE A SIGHT TO SEE] the whole of the heavens would be falling to one side and they would end up with a total REARANGMENT OF THE HEAVENS, [WHICH IS ANOTHER STATMENT FROM THE PAST.] On the opposite side of the earth THE SUN FELL FROM THE SKY [THAT IS WHAT THEY SEE].

## THE HEAVENS WERE TOING AND FROING

The heavens were to-ing and fro-ing, is a statement that is a follow on of the earth rolling over and rocking back and forth before settling properly into rotation spin, and then forms the new pole and new equator. If the world were badly out of balance, the pole would have moved quite some distance, forming a new equator as well, and would have to go through major earthquakes to reform its original and true shape. Some land bursting upwards and some land collapsing until balance is somewhere near correct. Some people actually lived through that.

-------------------------------------------------------------

## ANCIENT OBSERVATIONS AND TIME

All ancient observations have been right on the point of being totally DISMISSED, as being incorrect. -- THINK AGAIN. --Mathematics cannot be nearly right or NEARLY WRONG, as any person can correct nearly wrong by simple observations, at any future date. Haven't you noticed how we, of this time, have corrected observations from the past? They all seem to be slightly out. Look at our own calendar. Why have we added a leap year? They did not have it wrong; it was RIGHT for THEIR time. -- As I have written earlier, even weight has changed. I hope you now know, what I have been trying to make you understand.

As the orbit gets larger, it takes longer to get around the sun, and our position in our solar system is always changing, BUT IN OUR LIFETIME WE DONT SEE THE CHANGE, AS TIME TAKES SO LONG TO MAKE THAT CHANGE]]

## CHILDREN OF THE FUTURE

I write that earth picks up speed in rotation by maximum of four times, and any given weight is four times lighter. A 16 stone man would weigh only 4 stone. What about a newborn baby? A baby weighing 8 pounds in our time will only weigh 2 pounds in the future? Would they be on their feet much earlier, as weight as we know it is not there? That is balanced out by everything on earth slowly getting bigger and increasing bulk so as to stay on the ground.

----------------------------------------------------------------

## NEW ZEALAND SOUTH ISLAND RIVERS

South Island Rivers are very big rivers with a trickle of water running through them, WHY? There was obviously a lot of water in the past, as can be seen by simply looking.  Go back in time [[1000 years]] and look at the earth, it has just rolled over and out of orbit, then take a look at the position of the mountain range of the south Island. When the world rolls over, it does not stop turning, but it will be in a clockwise spin.  Now the sun, which was rising from the WEST and setting in the EAST will rise in the EAST and set in the WEST. The weather pattern will be changed.  The mountain range of the south Island will really make the rivers work.

Weather patterns will change and have to come across open country before hitting the mountain range. The water will come back to a greater land mass and the land will be more productive. The South Island will be the North Island, until the world rolls over again 2,000 years later, [1,000 years ago] In the Maori myths, the NORTH ISLAND IS WHAT THEY KNOW AS THE SOUTH ISLAND, AND IN THE SHAPE OF A GIANT STINGRAY FISH. When I was young I was told that pakiha simply don't know which way is up or down. That was as good as any explanation that I have heard, and one that seems to explain the difference in both our understanding of the way things are, as the Maoris say, SOUTH is UP and NORTH is DOWN.  We still have a lot to learn, don't we? 1.000 years ago earth rolled over, and the two Islands changed position again.
This is OUR time with the sun rising from the EAST and setting in the WEST. [OUR TIME, meaning the START of THIS NEW GENERATION, AFTER the flood and earthquake, catastrophe.)

----------------------------------------------------------------
## FREAK WAVE
[YEAR 2.000]

The world is turning too slow in rotation, so slow that the ozone is unable to be spread evenly, and too slow to spin on its own axis, as the world wobbles as it is turning.  Sometimes at sea there is a giant freak wave. When conditions are right on earth, the last to come into place is earth's wobble and the moons position. The moon has been following earths wobble and when the earth is down and begins to lift again, it has to pull the moon back as well, earths wobble is very slow but it leaves the moon behind and the distance between the earth and moon, means a small movement from earth, a much greater movement is forced on the moon and that is where all things stress out enough to pull at the water, and be able to raise a wall of water so high that any ships in that area will be lucky to survive  = FREAK WAVE. Scientists have dates and times of these GIANT WAVES, so it can easily be sorted out to get a picture.

-----------------------------------

## THE PLAUGES OF EGYPT

The earths cycle will bring on these problems when in the SEVEN YEARS OF FAMINE [No rain] For anybody to forecast any plagues; they would have knowledge of it happening before. The green belt that the Egyptians created with their copy of how an oasis works, [the pyramids] TO HEAT THE UNDERGROUND WATER, AND FORCE IT TO EXPAND, WHICH IN TURN FORCES THE WATER TO CLIMB TO THE SURFACE.] Which attracted locusts from all directions, and this green belt stands alone, food for living animals or insects. The priests in Egypt would have all this on record. Their records go back 13.000 years, so where are these records? THEY have seen FIVE SUNS appeared to have all died. One sun died every 2.000 years. OR SO THEY THOUGHT. IT HAD NOT RAINED FOR SEVEN YEARS AND ALL THE CLOUD ABOVE EARTH TRAPPED THE DUST FROM THE EARTHQUAKES, AS WELL AS THE FINE DUST FROM ALL THE VOLCANOS, IN TURN BLOCK OUT THE SUN. SO THEY SAY THE SUN DIED. Year 2000 is the seventh sun.  Only one answer, records are missing, but survive in MYTHS as explained earlier in my writing of part one.

-----------------------------------

## G O D

Who is GOD, Is he human? Is he millions and millions of years more advanced, compared to us?  "Let US make them in our own image" suggests there is more than one and do they understand DNA to a point where they cannot die?  Have they found a way to repair them selves? Is that what we call a GOD? BECAUSE THEY DO NOT DIE, AND DID THEY PUT US TOGETHER?  Is our god a jealous GOD? WOULD A GOD KILL PEOPLE WHEN HE HAS THE POWER TO SHOW THEM THE CORRECT WAY TO GO, WITH JUST A SIMPLE STATMENT OR A WAVE OF HIS HAND. A GOD that kills is a dangerous GOD. The GOD that I believe in will not interfere here. We have to LEARN for OURSELVES and we only have 2000 years of this generation to get it right. ALL being of like mind, with the phrase of, ALL FOR ONE AND ONE FOR ALL.  ALLREADY 1000 YEARS HAS GONE ONE THOUSAND TO GO. We will all have to work together to make it so.

-----------------------------------

## DISPLACEMENT OF WATER

I have wondered why seawater just quietly flooded land with no giant waves or storms. I have read this in many books and also seen TV documentaries on this subject. These floods did cover vast areas of high ground, then all the water went back to the sea, just as quietly. In a lot of places round the Earth, when the seawater left this higher ground, only puddles and wet ground and stranded fish and whales was left to show that this high ground was flooded with seawater.  Lets put all this into perspective and see what can be seen. Water that is on the surface of the Earth is in most of the lower levels. When the world rolled over [[for an insight we will say it rolled to the right]] the water stays where it is AS IT IS A LIQUID and the land underneath, will roll away leaving the water behind. The higher GROUND ON THE LEFT WILL PLUNGE ITSELF STRAIGHT INTO THE DISPLACED WATER.  The water will stay up on higher ground until the Earth completes its full roll over motion, and then will come back to the seabed, leaving sea creatures stranded.

-------------------------------------------

## M O S E S

God has given us free will for a reason. We are to experience life with all its ups and downs, and with all its frailties so as we can learn and make a choice, with our own experience. All in this new generation of 2000 years, before the last of this generation, gets involved with earth's catastrophe. 1000 of these years have gone already. I am convinced as to how Moses parted the Red sea. The water had to be on his right and left, stretching back behind him. When the world rolled over, it rolled the way he was facing and forced the water in front of him to part and move to left and right, leaving an open channel in front of him. We could work out how long the world took to roll over [[Joshua had two days in a row and on the other side, they had two nights in a row]] and the channel would stay open until the world was completely overturned. It looks like they gained time by resetting the suns position to restart the day. When that movement stopped, the water came back to its correct position [[as water does]] There would have been a lot less water in the red sea at that time, as it had not rained for seven years. [Their years, not OURS]  I had no idea that the RED SEA would look as I describe 1000 years ago, AND STILL BE THE SAME TODAY. There it IS, water either side of a peninsula as I describe, BUT, THAT IS THE ONLY WAY THAT A SEA CAN BE PARTED, SO LOGC THINKING WINS AGAIN.

----------------------------------------------------------------

## N O A H

Noah belonged to the last generation before the last catastrophe. They had 1,000 years moving out from the sun and 1,000 years moving back, they did not advance very much at all. After 2,000 years they were still riding horses and waving swords about, far too busy fighting each other to make any advancement. At that time they would all be fighting over food and water. We all know what happened to that generation, they nearly all died in the earthquakes and floods, only small bands of people survived to start the next generation {THATS WHO WE ARE] We have already moved out 1,000 years and we are involved with space [[at the year 2,000]] so within the next 1,000 years going back to the sun, I think we will be space and have the ability to leave at will. God left us here with free will, and that means no interference. So who was it that designed the ark and went away to bring all animals back to be saved on NOAHS ARK? Was it people who left here, the generation BEFORE NOAHS GENERATION, AND ACTUALLY CAME BACK TO HAVE A LOOK, AND TO SEE IF WE NEEDED HELP, OR WERE THEY FROM MUCH FURTHER BACK IN TIME AS THIS CATASTROPHY OF EARTHQUAKE, THEN FLOOD, HAS HAPPENED many TIMES IN THE PAST, ONCE EVERY 2.000 YEARS.   If people from the past did come back, how advanced would they be? Maybe if they or he were really advanced, so far ahead of the people who live here today, by just looking at the things they could do we just might think he is a GOD. (IS THE MEANING OF THE WORD GOD, "THEY WHO CANNOT DIE AND KNOW OF ALL THINGS"?)

-------------------------------------------------------

## DANIEL AND THE LIONS

Who was it that saved Daniel from the lions? A hard look at what might have happened surprised me, as it was so logical.   Lions use urine to mark their territory. Lions that were kept in a large cage, all together, will have to have a dominant male. The dominant male could not attack, or drive younger males away. He would have to have all other males submitting, if not, he would have to kill them. Someone knew lions quite well, he knew their ways, and used one of their own methods of control, against them.  I think that someone soaked Daniels clothing in the dominant males urine, let the clothing dry, and then had Daniel PUT THE CLOTHING BACK ON. HE MUST HAVE HAD INSTRUCTIONS AS WELL, SUCH AS " DO NOT RUN, STAND, AND DO NOT LOOK THEM IN THE EYES; LET THE LIONS COME TO YOU, LET THEM CATCH YOUR SCENT ". When the Romans called for the lions, ALL THE LIONS, EXEPT THE DOMINANT MALE WERE LET INTO THE ARENA. --- AS THE LIONS APPROACHED DANIAL, THEY RECOILED AWAY WHEN THE DOMINANT MALES URINE WAS DETECTED. To get anywhere near Daniel would be open to death by the dominant male. ------------ That would have to be one of the world's best-kept secret tricks, if that was, the way it was.

--------------------------------------------------------------------------------

## ONE-INCH SCALE

Winding a one-inch hose around a six-foot diameter drum, gave me the expected sixty feet in length. It looks like the Earth closes in on the sun sixty times its own width. [[As close as I can work out with the dates and time that I have got.]] Earth too far out from the sun will be too cold, and too close to the sun will be too hot, and BEYOND both extremes will be unable to sustain life as we know it. On the smaller orbit Methuselah was 969 years old, as time was too fast. We of THIS time [[year 2000]] have trouble making it to 100 years old

------------------------------------------------------------

## THE HUMAN BRAIN

Does the brain take a step forward without us knowing? Does a certain level of brain activity, force a new part of the brain to activate, which is a higher concentration of understanding. I think so. I will give an example. What made me look suddenly to my left, straight into the eyes of someone who was looking at me? I had no thought to look left, what made me look in that direction. Lets reverse the situation? I was at a counter in a shop working on H / P agreements, when the whole room went into shadow, I looked up at the front window and saw that a bus had parked outside, stopping the sunlight. Framed in the middle of the shop window was a young woman sitting in the bus, head and shoulders visible. I thought, "what a beautiful woman". She immediately looked left, straight into my eyes. She did not search for someone who was looking at her. That sudden turn left straight into a pair of eyes, looking. Why? It was as if she heard my thoughts, and I was way back in the shop, behind a counter. HER SUDDEN MOVE TOOK ME BY SUPRIZE AND I WENT STRAIGHT BACK TO MY PAPERWORK AS IF I HAD DONE SOMETHING WRONG. Is this the first hint of a new part of the brain being activated? People who have read this, all say "THAT HAS HAPPENED TO ME TO". There is something there, that at this time, cannot be easily, explained.

------------------------------------------

# PYRAMID

## BUILDING THEORY

The pyramid would be built flat, one layer at a time, full width. They could not take the large blocks in, so they simply placed them in the centre of the first floor, then lifted them one floor at a time until their position arrives, which then will be left open. The building will continue upwards leaving the opening, until the large blocks could be dropped into place from the top and then closed over as the building continues on upwards. The outer walls on one side only, will have a ramp starting from ground level, then on up to the first floor which will be used to build the second floor. The second floor will have the ramp continuing upwards from the first ramp so as to be able to build the third floor from the second side of the building. Each floor of the building will have a ramp built going up to the next floor around the outside of each of the four sides, climbing up all the way to the top with the ramp being the last section of each floor. When the top of the building is completed they simply backed down the ramp, back filling all the way down to the ground .The ceiling of the ramp would be blocks that are set into the inside walls and left to overhang, leaving an open channel following the ramps underneath, all the way to the top. Very little overhang is left as the four sides of the building lean inwards and blocks on top of the overhanging blocks lock up the structure. Don't forget, all these blocks of stone would be FOUR times lighter in their time, as the earth is in rotation spin, FOUR times FASTER.

## CONCLUSION

## MOON AND EARTH AT YEAR 2.000

For many thousands of years the moon has been swinging around earth, faster in orbit to earths own rotation speed. Looking back, I found that there was something missing. Everything I had on paper all made sense, but still, what was I missing? Something with ENORMOUS POWER. What was it? Again I read, and then again. Then a light of something --I was relying on writings and records from the past. INTERPRETATION was the key, that's the key word, interpretation. What if what they were seeing was not the cause of what was happening? MISINTERPRETATION.
There has been a lot of that already, so why not read again with that in mind, I now see what I was missing. See if you can pick up what I missed yourself by reading the first line of this page. Does the answer blast out at you? If not I will explain, [later.] We all know that the moon controls the tides, and the full moon with help from the sun, produces our higher tides. The moon also has helped to pull the earth out of shape by stretching the equator, and with earths own spin we end up with the best balance that the earth needs for orbit and rotation

spin. The plates of earth are moving very slowly over the surface, all caused by the moon and earths rotation.    Australia and the south Island of New Zealand are moving NORTH, TOWARDS THE EQUATOR, BOTH UNDER THIS STRESS, and LIKE OTHER PLACES AROUND THE EARTH. Are you beginning to see?   The earth moving away from the sun for one thousand years, with the moon helping to move the outer land mass, eventually UNBALLANCES THE EARTH.  The earth slows in rotation as well, and it's slowest, at the extreme distance from the sun in the year 2.000. A.D. The planets will soon line up with the sun which signals the time for all planets to go back TO THE SUN. All planet orbits will get smaller which makes the year shorter and shorter as time goes on. Earth is still doing the same orbit speed in the smaller orbit and that forces the smaller orbit to start overlapping. The seasons will come earlier and earlier as the orbit gets smaller.

The smaller orbit also forces the earth to increase its rotation speed and our scientists have picked that up. I gave the scientists 5 years to notice that the earth will start picking up revolution speed. [The year 2.000, to 2.003, rotation speed has picked up by point 26% in only three years and they have not yet found out the reason WHY?] They ALSO have not picked up that our orbit is slightly smaller, YET.  I am forced to back peddle here, back 300 years to the year 1700 A.D .I was not going to mention the mini ice age but the ozone still has to be mentioned. On the way out from the sun, the earth started to freeze. The ice came down from the North Pole and covered England, and a lot of people died from the cold.   [[I do remember seeing a painting of a young girl skating on the Thames, and all the ice that was about at that time, but had no idea really, what it was all about.]]  The earth had slowed in revolution, and OZONE COULD NOT BE SPREAD AROUND EARTH EVENLY AT THAT SLOWER SPEED. THE OZONE GOT UNEVEN, VERY THICK ON ONE SIDE AND VERY THIN ON THE SUNNY SIDE. THATS WHERE THE EARTH STARTED TO FREEZE, THE EARTH KEPT SLOWING IN REVOLUTION UNTILL THE VERY THIN SIDE OF THE OZONE SPLIT OPEN AND LET IN EXTRA LIGHT THAT BURST INSIDE THE OZONE CIRCLE TO HIT EARTH AND PASS EARTH, AND GO ON TO HIT THE OPPOSITE SIDE OF THE OZONE CIRCLE, ONLY TO BOUNCE AROUND,  [[no hole on this side, for the light to get out.]]   AND EVENTUALY BACK TO EARTH, CAUSING A HOT HOUSE EFFECT.  It would be like having TWO SUNS, No wonder the ice is starting to melt, and upset weather caused extreme temperatures at both ends of the scale, and will continue to do so until 2,300 when the earth will have picked up enough rotation speed to close the ozone hole. [[At the same position 300 years ago, when the ozone opened]]   WE ARE ALSO INVOLVED WITH ANOTHER PROBLEM. The ozone hole is on the very thin side and facing the sun. THE EARTHS LIFT IN ROTATION SPEED IS RIPPING THE OZONE HOLE WIDE OPEN AND THAT IS ALSO ADDING TO THE WEATHER EXTREAM. Lucky for us, the ozone hole is seasonal, as in New Zealand, summer is on the upside and winter is on the downside of earth, with its tilted rotation spin.  THE OZONE WILL EVENTUALLY START TO CLOSE WITH EARTHS ROTATIONAL SPEED AND

ANOTHER MINI ICEAGE WILL FOLLOW AS WE ARE STILL TOO FAR FROM THE SUN, EXACTLY THE SAME ORBIT AS IN THE YEAR 1,700.The moon has been pulling against earths hold, which for 1,000 years, [[In this new generation]] has been helping to move earth's plates, the extra rotation speed of earth will in time, refocus the earth's poles. [[AT THE YEAR 2.300, WHICH IS 300 YEARS FROM THE YEAR 2.000]] The frozen circle will move to one side as a new area is forced into the new pole, which is suddenly frozen as the new pole takes over. This is not an ice age; it is only the world rebalancing, as earth still has to pick up more speed. WITH THE EXTRA SPEED TO COME, THE EARTH WILL BE IN BALLANCE. Scientists will be able to work out what area will go into the new pole by noting which side of the earth is heavier and forcing the earth out of balance. This new area will go into the pole and should be vacated, BEFORE THE SHIFT HAPPENS, AS THE FREEZE WILL HAPPEN VERY QUICKLY [[ANIMALS WERE COUGHT IN THS SUDDEN FREEZE AND THEY ARE STILL FINDING MAMOTHS QUICK FROZEN IN THE ICE TODAY, THREE KILOMETRES UNDER ICE]] but that shift in pole position was a very big shift 13,000 years ago and we have had five separate small pole changes since then. The change in the pole will only be as much that is needed, to rebalance the earth. THIS ADDS TO THE ICE THAT IS ALLREADY THERE, AND MISTAKENLY THOUGHT TO BE AN ICEAGE, AS IT HAS DONE FOR MILLIONS OF YEARS IN THE PAST, CAUSING MASS EXTINCTION. Now that the earth is in balance it can reach the expected FOUR times rotation speed, quite safely, as I cannot find any more real big problems until the SEVEN YEARS OF PLENTY AND SEVEN YEARS OF FAMINE. With earthquakes, SUNAMI, and then volcanoes, that starts free flowing lava and blowing black poisoned soot into the heavy cloud. The heavy cloud eventually blocks out the sun. Seven years of plenty obviously means plenty of WATER AND HEAT [[TROPICAL]] Seven years of famine obviously means NO WATER AND EVEN MORE HEAT. [[DROUGHT]] It cannot rain as the clouds are too hot and are unable to condense and form rain. ALL THE WATER THAT IS TRAPPED IN THE AIR, ALONG WITH HELP FROM THE OZONE, TOGETHER ACT AS A VERY EFECTIVE BARRIER, PROTECTING THE EARTH FROM TOO MUCH DAMAGE. Now Earth will be circling the sun close to 10 times faster and earth's rotation is four times the speed that it is today, as Methuselah lived for 969 years, THEIR YEARS, NOT OURS. Everything, because of this higher rotation speed is four times lighter than it is today. They took their year to be from summer to summer, AS WE DO. THE SUN IS RESPONSABLE FOR PULLING ALL OUR PLANETS BACK AND LIFTING THEIR ROTATION SPEED. After seven years of no rain, the planets will now be all in line, and people on earth, in the final days of closing in on the sun, would see the sun coming down to earth and experience the violent earthquakes and floods that followed. [[And we think this report, is a myth]] THE SUNGOD CAME DOWN FROM THE HEAVENS TO DESTROY AND KILL". Misinterpretation is written down again. HERE IS WHERE THE LIGHT TURNED ON AND LET ME SEE, WITHOUT BEING THERE. As the orbit of all our planets is getting smaller and smaller, they are traveling less distance and as they close in on the sun they complete full orbit in less time, and keep on picking up more and more speed that reaches maximum speed just before a massive slingshot effect.

|        | OUR TIME 2000 | YEAR 3000    |
| ------ | ------------- | ------------ |
| EARTH  | 1040  m p h   | 10400 m p h  |
| MOON   | 2300  m p h   | 23000 m p h  |

IT IS THE MOON THAT PULLS AT EARTH AND CAUSES THE MASSIVE EARTHQUAKES. THIS EXTRA ROTATION SPEED HAS ALSO ALLOWED PEOPLE OF THAT TIME, TO APPEAR TO LIVE MUCH LONGER, AS THEIR TIME IS TOO FAST. THEY ONLY HAVE A THREE-HOUR DAY AND A THREE-HOUR NIGHT. MATHUSALA SUPPOSEDLY LIVED FOR 969 YEARS, ALONG WITH EVERYBODY ELSE AT THAT TIME AROUND EARTH WHO REPORTEDLY LIVED FROM AROUND 800 AND EVEN BEYOND IOOO YEARS OLD. THE EXTRA ROTATION SPEED ALSO FORCES THE DEEP WATER TO BULGE OUT FROM THE SEA BED AND PULL THE WATER AWAY FROM THE SHORE. AS ALL PLANETS LEAVE THEIR SMALL ORBITS THEY PULL OUT PARTS OF THE SUN WITH A SUDDEN BURST. ALL OF THIS MOLTEN ROCK, GAS AND FIRE, HURLES ITSELF STRAIGHT AT THE PLANETS AND ARE MISTAKENLY THOUGHT TO BE ASTEROIDS HITTING ONE PLANET AFTER ANOTHER. I AM UNSURE IF ALL THIS DEBRIS HITS OUR PLANETS. DOES SOME ACTUALLY BYPASS AND HEAD OUT INTO DEEP SPACE?

MASS EXTERMINATION STARTED WITH THE LACK OF RAIN, BEFORE GETTING TOO CLOSE TO THE SUN AND STILL CONTINUES FOR YEARS AFTER THE SLINGSHOT EFECT THAT ALL PLANETS WERE INVOLVED WITH SO AS TO ESCAPE FROM THE SUN. FIRE AND ROCK FALLING FROM THE HEAVENS JOIN IN ON THE EARTHQUAKE AND DROUGHT. THE SUN IS SEEN TO GO BACK TO THE HEAVENS. "DRIBBLING LIKE AN OLD MAN." MOST OF THE CLOUDS ABOVE EARTH TURN BLACK, DUST FROM THE EARTHQUAKES AND VOLCANIC ASH BLOCK OUT THE SUN AND THAT STATMENT IS BACK THE FRONT. EARTH WAS INVOLVED IN THE SLINGSHOT AND IT IS EARTH THAT IS MOVING AWAY FROM THE SUN. THEY WROTE DOWN WHAT THEY WERE SEEING [[MISINTERPRETATION AGAIN]] WHEN EARTH FINALLY REACHES ORBIT FAR ENOUGH FROM THE SUN, THE CLOUDS CAN CONDENCE AND IT BEGINS TO RAIN. THE WATER THAT FIRST FALLS IS GREETED WITH JOY FROM ALL SURVIVERS. BUT IT IS POISONED. THE WATER THAT WAS RAISED BY THE PYRAMIDS WILL BE FILTERED AND WILL POSSIBLY BE CLEAR OF THE POISEN. THE MAYAN WENT A STEP FURTHER BY ADDING LAYERS OF MICA [[ROCK]] UNDER THE FLOOR OF THE TEOTIHUACAN PYRAMID OF THE SUN, AND THE

MICA TEMPLE IN MEXICO. MICA HAS TRACE ELEMENTS THAT THE BODY NEEDS, SUCH AS LITHIUM, TITANIUM, POTASIUM, MANGANESE, MAGNESIUM, AND TWO TYPES OF IRON AND THE LIST GOES ON, ALL BENAFICIAL TO THE HUMAN BODY. WATER LEACHERS THE TRACE ELEMENTS FROM THE MICA. MAYAN TRAVELED OVER 2000 MILES TO FIND THE MICA THEY NEEDED, AND THEN TAKE IT BACK TO MEXICO. "WE HAVE FOUND WHERE THEY REMOVED THE MICA FROM THE GROUND, AND THESE ELEMENTS SHOULD BE IN OUR WATER TODAY." AFTER THE MASSIVE EARTHQUAKES AND THEN FLOODS, SURVIVERS UNDER MUCH STRESS, ARE FORCED TO START AGAIN AS MOST HUMAN IMPROVEMENTS ARE LOST. --- EVERYTHING HAS GONE, THERE IS NOWHERE THEY CAN GO TO BUY NESSESARY TOOLS, FOOD, CLOTHING, AND WITH NO COMUNICATION THE SURVIVING HUMANS HAVE TO START AGAIN. [[THE NEW GENERATION]]

-----------------------------------

## EARTHS OZONE, ICE AGE, AND MASS EXTINCTION, ON A GRAPH.

OZONE – the ages of human, going away and back to the sun.
(OZONE CLOSED in 300 to 400 years going back to the sun)

MASS EXTINCTION      [ICE GONE]]

[_____Heat_____]          (MINI-ICE-AGE)-<GOING BACK TO THE SUN

| 1000 | 900 | 800 | 700 | 600 | 500 | 400 | 300 | 200 | 100 |

(ALL WEATHER TO EXTREAM HOTHOUSE EFFECT, FROM I00 two300 years)

_____  _

GOING AWAY FROM THE SUN

OZONE OPENING AT 3OO YEARS

[HEAT……........................} >>   (ICE)

(Mini- ice age)

| 1000 | 900 | 800 | 700 | 600 | 500 | 400 | 300 | 200 | 100 |

-----------------------------------

FOCAST OF MAJOR PROBLEMS WE HAVE TO PREPARE FOR, AS WE GO BACK TO THE SUN.

All that you see in the above graph will be repeated on the way back to the sun, to understand these problems that we face. I will list them, as they will happen.

FROM YEAR 2.000, EARTH ON THE WAY BACK TO THE SUN WILL BE DOING A SMALLER AND SMALLER ORBIT, AND THAT WILL LIFT EARTHS ROTATION SPEED AND COMPLEAT THE ORBIT EARLIER. THERE IS A HOLE IN THE OZONE THAT WILL BE RIPPED OPEN BY EARTHS LIFT IN ROTATION. OZONE BEING VERY THIN AROUND THE HOLE WILL BE RIPPED RIGHT AROUND THE WESTERN SIDE OF EARTH.

THE HOTHOUSE EFFECT WILL CAUSE MUCH MORE DAMAGE AND THE WEATHER WILL GET MORE AND MORE EXTREME AT BOTHENDS OF THE SCALE

AROUND THE YEAR 2.300 THE OZONE WILL CLOSE, AS THE EARTH HAS PICKED UP ENOUGH SPEED TO PULL BACK THE OZONE. [[EARTH WILL BE IN THE SAME POSITION IN ORBIT THAT IT WAS WHEN ON THE WAY OUT AND TOO FAR FROM THE SUN, IN THE YEAR 1.700]]

AFTER THE OZONE CLOSES, WE WILL STILL BE TOO FAR FROM THE SUN, AND WE WILL HAVE ANOTHER MINI ICE-AGE [[THIS MINI ICE-AGE IS THE ONLY TRUE ICE-AGE, AS THERE IS ANOTHER REASON FOR ICE BEING IN THE WRONG PLACE]]

AROUND THE YEAR 2500 (500 FROM NOW, HALF THE WAY BACK TO THE SUN) EARTH HAS PICKED UP TWICE THE SPEED THAT IT IS DOING NOW, IT IS OUT OF BALLANCE AND EARTH DOES AN AMAZING REBALLANCE. ALL THE PLATES HAVE MOVED AND FORCED EARTH TO WOBBLE. BOTH POLLS SUDDENLY REFOCAS THAT LEAVES ICE DOWN THE SIDE OF EARTH AND GOES ON TO FORM TWO NEW POLLS, CATCHING WHOMEVER BY TOTAL SUPRIZE. FREEZING EVERYTHING ON THE SPOT.THIS CAUSES MASS EXTINCTION OF ANIMALS. [[AS YOU NOW SEE, THIS IS NOT A TRUE ICEAGE, IT IS ONLY THE EARTH REBALLANCING]]

AROUND THE YEAR 1900 [[2900YEARS FROM NOW IN OUR TIME]] WE ARE GETTING TOO CLOSE TO THE SUN. THE WHOLE EARTH TURNS TROPICAL, AND

STAYS TROPICAL FOR SEVEN YEARS. [[IT WAS A GOOD TIME FOR DINOSAURS, AND AS THE WORLD IS TURNING IN ROTATION NEARLY FOUR TIMES FASTER, THEIR WEIGHT IS VERY MUCH LIGHTER,]] AND THERE IS SO MUCH FOOD.

THE LAST FOURTEEN YEARS, CLOSING IN ON THE SUN GETS VISUALLY STUNNING. THE PLANETS HAVE LINED UP AND ARE SWINGING AS A UNIT AROUND THE SUN, CLOSE TO SLINGSHOT SPEED AND THIS IS THE MOST TROPICAL THAT EARTH GETS. [[SEVEN YEARS OF PLENTY]]

THE LAST SEVEN YEARS [[SEVEN YEARS OF FAMINE]] EARTH IS TOO CLOSE TO THE SUN AND IS SO VERY HOT, THAT THE CLOUDS ABOVE EARTH CANNOT CONDENCE AND FORM RAIN. [[THIS IS WHY THE FAMINE]] EVERYTHING SHUTS DOWN, NO FOOD FOR EVERY LIVING THING. THE EGYPTIANS WERE EXPECTING THIS, AS THEY HAD IT ALL ON RECORD AS HAPPING FIVE TIMES BEFORE AND HAD MADE SURE THAT THEY COULD GET WATER BY BUILDING PYRAMIDS WHICH ARE A COPY OF HOW ALL OASIS WORK. BY SIMPLY HEATING THE UNDEWRGROUND WATER AND FORCING IT TO RISE TO THE SURFACE. WE ALL KNOW THAT BY HEATING WATER IT EXPANDS (JUST TAKE THE TOP OFF A HOT RADIATOR AND WATCH THE EXPANDED WATER BLOW OUT UNDER PRESSURE. IT IS NOT BOILING AT THAT STAGE, IT IS ONLY UNDER PRESSURE) WHY HAVE WE MISSED THIS VERY IMPORTANT POINT? AS I HAVE WRITTEN EARLIER, THE PYRAMIDS WERE MUCH EASIER TO BUILD IN THE LAST YEARS CLOSING IN ON THE SUN. THE EARTH WAS TURNING CLOSE TO FOUR TIMES FASTER AT THAT STAGE, WHICH MAKES EVERYTHING FOUR TIMES L- I- G- H-T-E-R.

NOW FOR THE LAST MOVEMENT OF ALL OUR PLANETS AS THEY CLOSE IN ON THE SUN, WHICH ARE INTERPRETED INCORRECTLY FROM OUR PAST. I HAVE EXPLAINED THAT EARLIER IN MY WRITTING SO I DONT NEED TO GO INTO SO MUCH DETAIL SO I WILL EXPLAIN IT AS IT ACTUALLY HAPPENS. ------- ALL INLINE PLANETS ARE COMPLEATING FULL ORBIT FAR TOO QUICKLY AT THIS POINT NOW AND THE FINAL DIVE STRAIGHT AT THE CONSTANT PULL FROM THE SUN MAKES IT APPEAR TO MAN AS IF THE SUN IS COMING DOWN TO EARTH. IF THE PLANETS DID NOT HAVE WEIGHT AND SPEED TO OVERCOME THE SUNS MAGNETIC PULL, ALL THE PLANETS WILL DIVE STRAIGHT BACK INTO THE SUN. MASSIVE EARTHQUACKS HIT EARTH, VOLCANOS ERUPT AND BLOW, LAVA FLOWS, THE WHOLE EARTH SHUDDERS. THE NOISE IS DEAFINING. THE GROUND SPLITS AND UNDERGROUND WATER BURSTS OUT HIGH INTO THE AIR. THE GROUND IS LIFTED THREE KILOMETRES INTO THE AIR

AND ROLLS OVER ITSELF. (WE HAD BETTER BE IN SPACE AT WILL BY THEN, AS FAR TOO MANY PEOPLE WILL DIE, ALL CAUSED BY THE SUN AND THE MOON) BUT, EVERYTHING IS TOO MUCH FOR THE LIQUID SUN AND THE PLANETS WEIGHT AND SPEED PULL A MASSIVE BALL OUT OF THE SUN AS THEY SLINGSHOT AWAY. THIS MASSIVE BALL FOLLOWS THE OUTWARD MOVING PLANETS AND HITS THEM ONE BY ONE AS THEY ARE STILL ROUGHLY INLINE. ITS THIS ROCK AND FIRE THAT HIT EARTH AND DESTROYED TWO CITYS IN THE BIBLE. BY THE TIME IT REACHS EARTH, THE SPLASH OFF FROM MURCURY HAS HAD TIME TO START COOLING. " ROCK AND FIRE FELL FROM THE SKY." THE CLOSEST PLANET GETS A FULL DIRECT HIT OF LIQUID SUN, WHICH FINALLY COOLS WITH TIME TO BECOME PART OF THAT PLANET. (WHICH NOW WILL BE BIGGER THAN IT WAS BEFORE, AND NEXT TIME WILL BE BIGGER AGAIN) -------- (ALL PLANETS COME FROM THE SUN AND WHEN THE CLOSEST PLANET TO THE SUN IS BIG ENOUGH, ALL PLANETS MOVE OUT ONE THEN STARTS ANOTHER PLANET. THE SUN IS THEN SEEN FROM EARTH TO GO BACK TO THE HEAVENS. (ALL THIS TIME, ANIMALS HAVE BEEN DYING THROUGH LACK OF WATER AND FOOD) AS PLANETS MOVE OUT, ALL THE CLOUDS ABOVE EARTH COOL ENOUGH TO CONDENCE AND THE SKY TURNS BLACK. (THE EJYPTIONS THINK THAT THE SUN DIED AS NO LIGHT CAN PENERTRATE THESE BLACK CLOUDS) FINALLY IT RAINS, AND KEEPS ON RAINING, YOU CAN SEE WHY IT RAINS BLACK RAIN NOW, AS ALL THE DUST AND SOOT FROM THE VOLCANOS COMES BACK WITH THE WATER, POISINOUS FOR ANY LIVING THING TO DRINK. "TO QUOTE" ------ ("WHEN ALL THE WATER WENT BACK TO ITS BED, THINGS FINALLY RETURNED TO NORMAL.) THE SURVIVERS THEN STARTED THE NEXT GENERATION. [[2000 YEARS]]

## NOTE FROM THE WRITER

EVERY BIT OF THIS WRITING WAS REINTERPRETED HISTORY, IT WAS LIKE A HUGE PUZZLE THAT COULD NOT BE PUT TOGETHER, SO I HAD TO GO BACK IN TIME TO THE OLD BOOKS AND INLCUDE AS MANY MYTHS AS I COULD RECOGNISE. THERE WAS SO MANY THINGS LIKE. "HE DROVE A CHARIOT THAT HE COULD NOT CONTROLE ACROSS THE SKY.") --- WAS THAT AN EXPLANATION OF THUNDER ROLLING ACROSS THE SKY?

"THE SUN WAS GOING TOO FAST"
"IT WILL APPEAR TO MAN AS IF THE SUN STOPS"
"MOSSES PARTS THE RED SEA"
"ROCK AND FIRE FELL FROM SKY"
"THE HEAVENS ARE TOING AND FROING"

"THE SUN IS ZIG ZAGING ACROSS THE SKY LIKE A FIREY COMET"
"THE SUN CAME DOWN TO DESTROY AND KILL ALL HUMANS"
"THE SUN WENT BACK TO THE HEAVENS DRIBBLING LIKE AN OLD MAN"
"WE HAVE SEEN FIVE SUNS DIE"
"THE SUN WE HAVE NOW IS THE SIXTH SUN" (12000 YEARS ON RECORD)
"THE HEAVENS FELL FROM THE SKY, WE ENDED UP WITH "NEW HEAVENS"
"THE SUN FELL FROM THE SKY"
"THE MOON FELL FROM THE SKY"
"GODS CAME TO VISIT, TEACH US AND HAVE CHILDREN WITH OUR WOMEN."
"THE SEA FLOODED VAST AREAS, THEN JUST QUIETLY WENT BACK"
"THE PYRAMIDS"
"AKHENATEN"
"THE DRAMATIC RISE AND FALL OF OUR OCEANS"
"CITYS UNDER THE OCEANS"
"CITYS UNDER GROUND"
"LOST CIVILATIONS"
"ICE ON MARS"
"AZTECS"
"WHY SACRIFICE [[BLOOD]] TO THE SUNGOD"
"EGYPTIONS"
"MATHUSALA AND EVERYBODY LIVING NINE HUNDRED YEARS AND OVER"
"THE SUN IS NOW RISING, WHERE IT WAS ONCE SETTING. "

AS YOU CAN SEE, THE LIST IS GOING TO GET TOO LONG,
AND WITH REINTERPRETATION OF HISTORY,
THE WHOLE LIST IS ALL THERE.


I   HOPE THE READER CAN NOW SEE RIGHT BACK INTO THE PAST AND MAKE
SENSE OF IT ALL.


REG GRIFFIN.